Crossroads:
History of Science,
History of Art

Essays by David Speiser, Volume II

Kim Williams
Editor

 Birkhäuser

Editor
Kim Williams
Kim Williams Books
Corso Regina Margherita, 72
10153 Torino
Italy
kwb@kimwilliamsbooks.com

2010 Mathematics Subject Classification: 00A66, 00A67, 01A85

ISBN 978-3-0348-0138-6   e-ISBN 978-3-0348-0139-3
DOI 10.1007/978-3-0348-0139-3

Library of Congress Control Number: 2011931276

Cover illustration: Force diagram by Sir Isaac Newton, Philosophiae Naturalis Principia Mathematica, first edition (1687): Book I, Sect. II, Prop. I, Theorema I (p. 37).

Cover photo: Baptistery of Pisa, by Kim Williams

Cover design: deblik, Berlin

Printed on acid-free paper

Springer Basel AG is part of Springer Science+Business Media

www.birkhauser-science.com

Contents

Foreword

Frans Cerulus

Perusing the titles of the essays in this book makes it clear that its author is a man with many interests and a great curiosity. David Speiser is a lover and connoisseur of art. His view of the world is coloured by his familiarity with mathematics, that is, with "organized imagination" (his definition, see p. 49). As a historian of physics and mathematics, editor of works of Euler and the Bernoullis, he lives part-time in the eighteenth century, from which he inherited an old-world *politesse*.

To me – besides a warm friend and colleague – David Speiser is first and foremost a theoretical physicist who contributed decisively to our understanding of elementary particles by discovering their fundamental symmetry and its effect on their decay. He knows first-hand how fundamental research is carried out. He was also for many years the professor who introduced the students at the University of Louvain to the beauty of classical mechanics and the theory of relativity. It led him to think deeply about the foundations of these sciences and his thoughts were valued by the many colleagues who invited him to share his insights through talks he gave in several of the languages he is familiar with. The present volume presents a selection of those talks to an English-speaking public. We should be especially grateful to Mrs Williams, who made the translations and edited the texts with great competence.

The first three essays are on works of art, but works where mathematics, hidden for the casual beholder, raises art to its highest level. Medallions of the sixteenth century B.C., hence long before the recorded history of geometry, reveal that the Myceneans knew a theorem on the geometry of the circle. Raphael, the master of Italian Renaissance painting, put on canvas the perfect likenesses of a rectangle viewed obliquely and an octagon viewed from different elevations, strictly geometrical problems never solved that perfectly before, scarcely noticed by art critics, but revealed here as an intellectual feat. David Speiser has special ties with Pisa and could not help but look at its monuments with the eye of the expert in group theory: he takes us on a tour of the Piazza dei Miracoli and shows how the builders were guided by uncommon symmetries. Like a detective, he relates the symmetry clash in the Baptistery – you will not find it in the guidebooks – to the enthusiasm that seized the Pisans in the thirteenth century for the newly discovered symmetries from Arabia.

The elation of the scientist upon making a discovery and the exultation of the artist are not unlike. Beauty does not dwell only in productions of the fine arts: a proof can be beautiful, a text can be admirably organised, and an experiment can be brilliantly conceived. In "What Can the Historian of Science Learn from the Historian of the Fine Arts?" our author meditates on these similarities, how they are perceived by the outsider and transmitted by the historian. History is not written once and for all, especially not in science: there are aspects of a theory missed by the contemporaries that shine surprisingly brightly in the light of later discoveries. How to make contact with the great minds of the past, how to distil the essence of their insights from the sometimes antiquated formalism and the conventions of the time, is the calling of the historian of science. In "Publishing Complete Works of the Great Scientists: An International Undertaking" David Speiser and his co-author, Patricia Radelet-de Grave, give very practical directions for the first step in the process: publishing, that is, making available and accessible, the original texts hidden in rare old journals ensconced in the vaults of a few libraries.

"Clifford A. Truesdell's Contributions to the Euler and the Bernoulli Editions" presents an ideal image of history of science in the way Clifford Truesdell made part of the heritage of Euler and Bernoulli come to life. The harassed teacher of mechanics who cannot possibly make time to read Truesdell's two volumes on the history of mechanics will find in David Speiser's essay a most lucid and useable compendium on the origin of classical mechanics.

In "*Gruppentheorie und Quantenmechanik*: The Book and its Position in Weyl's Work" we are told of an almost miraculous coincidence: the great mathematician conceiving precisely the mathematical structures that were needed almost immediately after to explain the essence of the new quantum mechanics, which led to a remarkable textbook.

But what exactly is it that makes a textbook remarkable? What is it that we should teach the young generation, and how should we do it? Certainly not as it is too often done, by power of tradition and intellectual laziness. David Speiser makes it clear that the ideal teacher should put his or her subject matter through a process of "digestion." As a historian and a teacher, he knows that the historical development is usually not the proper way to introduce the principles of a science. One should aim for simplicity, but uncovering the simplicity of a theory is a demanding task of "digestion" (see p. 146). One should concentrate on the key concepts of the theory. Why this is so, exactly what a concept is, how it relates to the mathematical description and the experimental facts: these are the core ideas of "The Importance of Concepts for Science," which is illustrated by the role of "force" as the central concept in Newton's mechanics.

Invariably, the physicist will meet the philosopher, because as soon as questions of method and meaning are raised the physicist steps outside the range of science proper. We might think that the matter surrounding us is real, in an intuitive sense. But the physicist knows that this matter consists mostly of protons and neutrons; pursuing ever further the concept of proton, he has to leave behind the realistic image of the small charged mass point and he may wind up (in a particular theory) with a very abstract notion: a singularity of a function of complex numbers. I cite this as a small alternate example for the essay "The History of the Sciences at the Crossroads of the Pathways towards Philosophy and History," which goes deeply into the relation of physics and philosophy, but whose final aim is to define more precisely the key role of the history of science in the confrontation.

Finally, "Remarks on Space and Time" shows how great minds of the distant and not so distant past – Newton, Euler and Leibniz, as well as Einstein and Minkowski – battled with the new ideas they had conceived and how it took two hundred years before Newton was fully understood. Even now we do not completely understand space and time. We are not justified in looking at space as a geometric abstraction: it is a physically active reality. Speiser, earlier on, had an intriguing but unanswerable question: how much of yet unknown mathematics will future physicists need in order to understand the world we live in? In what direction should we look? In 1980 he risked a look ahead: to understand the ultimate structure of matter we will need to rethink space and time. Thirty years later a sizeable number of men and women are trying hard to do just that, publishing on parallel worlds in ten dimensions and on time as an emergent phenomenon; we have clearly not reached the end.

There is, however, more in the book than a brief overview can make evident. All those essays proceed from talks where the speaker, now and then, follows a sidetrack by associative thinking. And because David Speiser read an impressive number of the classics in German, French, English, Italian and Latin, this makes for unexpected sparks.

History of science is a wide field. The most popular texts in this discipline are the short notices in textbooks, usually in a textbox and adorned with a picture of the great man. As a rule they are adapted from previous textbooks and constitute a subvariety of popular mythology. Then there are the biographies, often professionally written after extensive research; they describe the family life of the scientist, his position at court, his quarrels, his social network, but remain vague on the exact meaning and the significance of his discoveries. Finally there are the few histories by authors who are thoroughly familiar with the science of today, but are still able to inhabit a period of the past and have a passion to understand how their fellow scientists of that epoch were thinking. For those authors a kind of miracle can happen: out of the scrutinized and meditated old print emerges the figure of a living man, with his thoughts, hopes and illusions, his insights and struggles. As David Speiser recounts, André Weil expressed it admirably: "…I think I know him [Euler] better than I know most of my best friends" (p. 45). In his "Crossroads" essay David Speiser has given a list of such authors, and I will decidedly add his name to it.

Looking back at all the essays I discern a leading thread: to relive the emotion of new understanding, to know intimately the great men of the past in their moments of inspiration that shaped our thinking. David Speiser invites us into his study and we listen in on his conversations with the spirits of the great men who come to life in the span of an essay.

Leuven, January 2011

Kim Williams

It is a great pleasure for me to present this second volume of essays by David Speiser. This follows the first volume, entitled *Discovering the Principles of Mechanics 1600–1800*, published by Birkhäuser in 2008. As editor, this present collection is closer to my own interests and field of research, that is, relationships between the arts and the sciences, beginning as it does with the history of art and architecture, and then broadening into an examination of the similarities and differences between history of the arts and history of the sciences. One of these essays in particular is close to my heart as well as to my interests, because it is thanks to David's study of the symmetries of the Baptistery of Pisa, which led him to come across my own study of the Baptistery of Florence, that he first contacted me in 1994, thus beginning what grew into a very special friendship.

My interventions as editor of this collection include translating four of the papers from the original French and Italian, and correcting the minor errors that had crept here and there into the original publications. The author of the essays being a formidable editor himself, this was not a task I undertook lightly. As a service to the reader, all references have been checked, verified, and updated where necessary. I was much helped by Frans Cerulus, whose involvement in this volume went far beyond the writing of the Foreword, as he too proofread and corrected all the essays, especially the translations, found references which I was unable to locate, and made valuable suggestions regarding editing criteria. I am most grateful for his enthusiastic collaboration.

Few professionals – be they scientists, historians of science, or historians of the arts – can match David's breadth of interests and range of knowledge. Through these papers, most of which originated as lectures, and in which those who have heard the author speak will recognize his unforgettable voice, we meet a very large and varied society of characters. To aid the reader, it was decided to enrich the index of names with brief identifiers of the historical figures: nationality, field and dates. These identifiers are intended merely to provide clues, and are necessarily abbreviated, but will nevertheless be helpful for those reading outside their field. Just skimming the index will give the reader an idea of the vastness of the world into which these essays will lead him. It is a special collection indeed: a look behind the scenes at the creation of great works of art, the process from discovery to systemization to unification of great fields of science, carefully meditated discussions of interactions between science, philosophy and history. It is a privilege and an honor to enter into this world.

I would like to express my gratitude to David Speiser for entrusting this work to me; to his wife Ruth Speiser for her support and affection; to Frans Cerulus for his insightful Foreword and collaboration; to Patricia Radelet-de Grave for allowing me to include the essay she co-authored, and for answering questions and providing information and references; to Laura Garbolino of the Biblioteca "G. Peano" of the

Department of Mathematics of the University of Turin for help in obtaining the title pages of the books by Hermann Weyl and Clifford Truesdell; to the Director of the Vatican Museums, Dr. Antonio Paolucci, for permission to reproduce Perugino's *Delivery of the Keys to St. Peter* from the Sistine Chapel and Raphael's *Crowning of the Virgin* from the Pinacoteca Vaticano; and the Italian Ministery for Cultural Goods and Activities for permission to reproduce Raphael's *Marriage of the Virgin* from the Pinacoteca di Brera, Milan; to curator Caroline Joubert of the Musée des Beaux-Arts de Caen for permission to reproduce Perugino's *Marriage of the Virgin*; to Birkhäuser for support of this project, especially Thomas Hempfling and Karin Neidhart.

I hope the reader will enjoy these essays as much as I enjoyed working with them.

Torino, January 2011

Introduction

It is well known that many mathematical theorems were grasped intuitively long before the systematic development of mathematics. As is known, geometry was developed systematically by the Greeks of the fourth and third centuries B.C., but long before that, in Egypt, we find ornaments that achieve complicated symmetries, that is, those based on an entirely non-trivial group.[2] There appears to be a case of an ornament inspired by a geometric theorem among the treasures found in the shaft graves of Mycenae. It appears that up to now the Mycenaean culture has barely been studied from the point of view of the history of the sciences; this small note seems therefore warranted.

The Mycenaen Jewel

The National Archaeological Museum in Athens houses the large treasure of gold objects that Heinrich Schliemann discovered in Mycenae in 1876, in the shaft graves that date from the sixteenth century B.C. (today designated as grave circle A). From tomb III, he exhumed three female skeletons covered with a large quantity of gold jewelry. Notable among these are a number of discs in gold leaf measuring about 6 cm in diameter, some of which feature geometricized figures of flowers, butterflies, octopi, etc. On others, one finds geometric figures, and one of these medallions presents a theorem that every reader learned long ago at school. This appeared as medallion 20 in Georg Karo's fine work *Die Schachtgräber von Mykenai* (fig. 1).

Karo describes the object thus:

> 20. Pl. XXVIII. Small gold discs with star pattern. Diam. 6.2 [cm] …
> 60 exemplars, 16 with rather large, roughly punched holes. A six-pointed framework is produced from flat arcs, which is filled by a 6-pointed star, in its turn formed by 6 overlapping arcs. Between the leaves small circles are situated, with recessed insides. In 3 exemplars, the pointed leaves are filled with an engraved fishbone pattern …; one of these and another one are made out of lighter, thinner sheet metal.[3]

[1] Originally published as "La symétrie sur un bijou du trésor de Mycènes," *Annali dell'Istituto e Museo di Storia della Scienza di Firenze*, Anno I, Fascicolo 20, 1976.
[2] On this subject, see Andreas Speiser, *Theorie der Gruppen von endlicher Ordnung*, 4th ed. (Basel und Stuttgart: Birkhäuser Verlag, 1956) and Hermann Weyl, *Symmetry* (Princeton: Princeton University Press, 1952).
[3] "*20. Taf. XXVIII. Goldplättchen mit Sternmuster. Dm. 6,2 ….*
60 Exemplare, 16 mit ziemlich grossem, roh eingeschlagenem Loch. Aus flachen Kreisbögen ist ein sechsspitziger Rahmen hergestellt, der, wiederum durch 6 sich überschneidende Kreisbögen, mit einem Stern aus 6 spitzen. Blättern gefüllt ist. Zwischen den Blättern sitzen

Fig. 1. Mycenaean medallion.
From George Karo, *Die Schachtgräber von Mykenai*, vol. II, pl. XXVIII

This medallion was also reproduced by Giovanni Becatti in *Oreficerie Antiche*, where he describes it thus:

> Circular disc in gold stamped with a rosette with six petals inside a hexagon with curved sides, and small circles between the petals. From tomb III of Mycenae. Diam. 6.2 cm. Second half of the 16th century B.C.[4]

The essential characteristic of this figure is the fact that *all six* interior arcs have been drawn *with the same compass opening* that determined the big circle, i.e., the perimeter of the medallion (fig. 2).

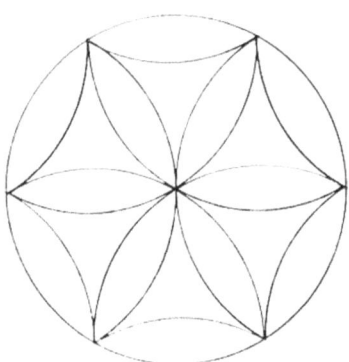

Fig. 2.

---

kleine Kreisen mit eingetieftem Innern. Bei 3 Exemplaren sind die spitzen Blättern mit eingeritztem Fischgräternmuster gefüllt ...; eines von diesen und ein anderes besteben aus hellerem, dünnerem Blech (Georg Karo, *Die Schachtgräber von Mykenai*, vol. 1 (text) and vol. II (plates), Munich: F. Bruckman AC, 1930, p. 47, fig. 20).

[4] "*Bratta aurea circolare stampigliata con rosette a sei petali entro un esagono dai lati curvi, e cerchietti fra i petali. Dalla tomba III di Micene. Diam. cm. 6,2. Seconda metà del XVI sec. a.C.*" (Giovanni Becatti, *Oreficerie Antiche*, Rome: Istituto Poligrafico dello Stato, Libreria dello Stato, 1955, p. 154, tav. XVIII, fig. 62).

This construction is based on the following theorem:

The radius *r* of the regular hexagon is equal to each of its sides.

Or, inversely,

> If one starts at any given point on a circle, and determines a second point on the circle as far from the first as it is from the center, and then repeats this process from the last point determined and continues in this manner, one will return, after the sixth point, *exactly* to the initial point (fig. 3).

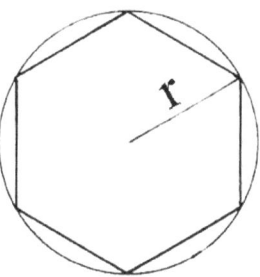

Fig. 3.

Many of the readers of these lines will remember the pleasure that they experienced when, for the first time, they understood this theorem, when they drew and redrew these circles with their first schoolboy's compass, and when the figures took shape in front of their eyes. There is no reason to believe that it was any different in ancient Mycenae, and we think that the *intuitive discovery of this theorem prompted the ornamentation of these medallions.*

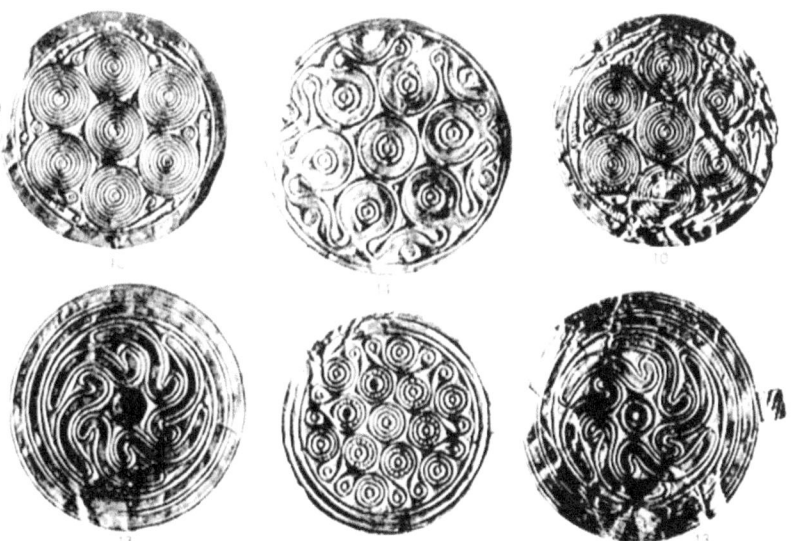

Fig. 4. Six Mycenaean medallions. From Karo, *op. cit.*, pl. XXIV

Karo had already suspected the possibility that the goldsmith who made these discs knew this theorem. Yet, he expresses this hypothesis about another ornament, shown in the upper right hand corner, numbered 10, in Karo's plate XXIV (fig. 4).

Like all the other disks reproduced in our fig. 4, this one too is the fruit of the discovery of this theorem; in our opinion, however, it does not provide the proof of it. On the other hand, the description of medallion 20, "Admittedly, the star consisting of six large arcs drawn from the rim with connecting flat arcs between the points remains an isolated case"[5] (note that ornament 20, XXVIII = ornament 81, XXXIV), in suggesting a distinction between "large arcs" and "flat arcs," probably does not do justice to the ornament. Note that we say: probably. Some of the exterior arcs, *but not all*, are indeed flatter; however, we believe that these inequalities are due to an imprecision in the execution rather than to the artist's intention. The fact that the twelve arcs of the ornament can be drawn with the same opening of the compass reinforces our thesis.

This procedure can be continued in all directions. From every pair of neighboring points on the circle, one can determine the exterior point equidistant from each of them. This point will then be the center of the concave arc (exterior) of the ornament, which is the starting point of another circle, and so on. Thus the medallion is only one part of an infinite ornament (fig. 5).

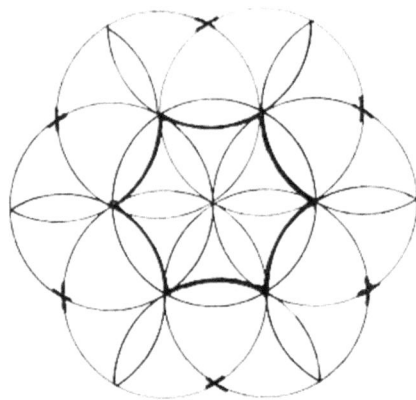

Fig. 5.

Still, one might object that it is not necessary to read the ornament in this way, and maintain that it merely depicts a geometricized star or a six-petaled flower, which are quite often found in Mediterranean art of this period. In response to this criticism, we put forward the following three observations:

- This interpretation doesn't do justice to the concave exterior circles;
- If the figure were a flower, we would expect to find similar flowers with 4, 5, 7, 8 ... "petals." It is true that a few examples of these are found in the Mycenaean treasure ; for example, on the large diadems found in shaft grave III, one finds some figures with "petals," but the great majority of the

---

[5] *Zwar bleibt Stern der aus sechs vom Rande aus geschlagenen grossen Bögen, mit verbindenden flachen Bögen zwischen den Spitzen, ein vereinzelter Fall* (Karo, *op. cit.*, p. 269).

ornaments have six or even twelve petals. Of the discs, the ornament of the figure 20 seems to be unique in its genre;

- Karo says of Mycenaean art that "the basis of the whole ornamentation is built on circles, wavy lines and spirals, which gives this art its stylistic character."[6] He tried to have the artist K. Grundmann reconstruct the genesis of some of these ornaments,[7] in order to show how the goldsmiths had arrived at these refined drawings, which he calls, rightly so, "quite complicated."

We have underlined the geometric aspect of medallion 20. This doesn't exclude the attribution of a symbolic significance, or even one that is religious or magical. In this regard, it is interesting to note that two of these medallions, also found in tomb III, were used as the two pans of a scale (fig. 6).[8]

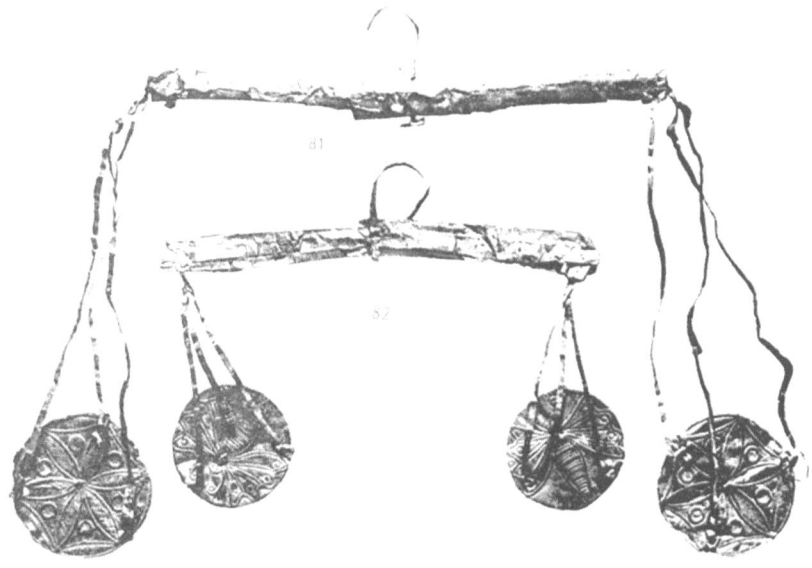

Fig. 6.

As this ornament is based on a precise theorem, it is a perfect expression of the idea of exact distribution and payment. Was it a symbol for this? Yves Duhoux communicated to me that it has been hypothesized that these scales may indicate a belief in the weighing of souls. Our remark may therefore provide support for this hypothesis.

Lessons for the history of the sciences

What lessons does this ornament hold for the history of the sciences?

First of all, that the use of the compass was well known in sixteenth-century B.C. Mycenae. Prof. B. L. van der Waerden drew my attention to the fact that the use of

---

[6] *Die Grundlage der ganzen Verzierung bilden Kreis, Wellenlinie, Spirale, sie geben dieser Kunst ihr stilistisches Gepräge* (Karo, *op. cit.*, p. 259).

[7] Cf. Karo, *op. cit.*, p. 265.

[8] See Oscar Montelius, *La Grèce Préclassique* (Stockholm, var Haeggströms Boktryckeri A.B., 1924), 1re partie, fasc. I, pp. 18–218. See also G. Karo, *op. cit*, vol. II, pl. XXXIV.

the compass is attested to in ancient Babylonia of the time of King Hammurabi (eighteenth century B.C.), by showing three circles drawn *with precision.*[9]

This comparison puts the discovery of the Mycenaeans in relief. For the Babylonians, the ratio between the perimeter and the diameter of the circle was 3:1,[10] whereas, in the Mycenaean figure, it is sufficient to draw the six arcs to see at once that this value is too small (fig. 7): the circumference of the circle is obviously larger than that of the hexagon, which, according to this figure, is equal to three times the diameter.

Did the Mycenaeans grasp this?

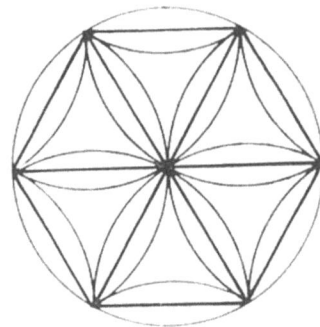

Fig. 7.

Further, the ornament shows that the Mycenaeans not only had a clear idea of geometric drawings and studied them systematically, but also that a theorem of the geometry of the circle was familiar to them, as well as a construction with the aid of a compass.

Questions for further investigation

We would very much like to have answers to both of the following questions:

1. Where did these goldsmiths get their geometric knowledge: did they discover this ornament for themselves, or did they learn it from somewhere else?

One is tempted to look for the source of this ornament in Minoan art. Indeed, some ornaments found in Crete – for example, the geometric ornament on the scepter in the shape of leopard[11] (somewhat earlier) or the ornament in spiral on the pithos[12] of the same period – would seem to lend weight to this hypothesis.

But Minoan art doesn't seem to have shared the "circular obsession" of Mycenaean art. In any case, a superficial examination didn't provide any tangible evidence, and the particular character of this period of Mycenaean art has often been underlined.

---

[9] See B. L. Van Der Waerden, *Erwachende Wissenschaft* (Basel und Stuttgart: Birkhaüser Verlag, 1956), fig. 20, p. 109.
[10] See B. L. Van Der Waerden, *op. cit.*, p. 120.
[11] Spyridon Marinatos and Max Hirmer, *Kreta und das Mykenische Hellas* (Munich: Hirmer Verlag, 1959), figs. 68, 80.
[12] Ibid.

2. Can we find subsequent traces coming from this ornament that show us that its mathematical content had been conserved and transmitted?

All that we were able to find is a clasp, found in Boeotia, reproduced in Erwin Bielefeld,[13] which, according to a friendly communication from Yves Duhoux, dates from the first third of the last millennium B.C (Fig. 8). The ornament is an extension of the ornament of Mycenae (compare to our fig. 5).

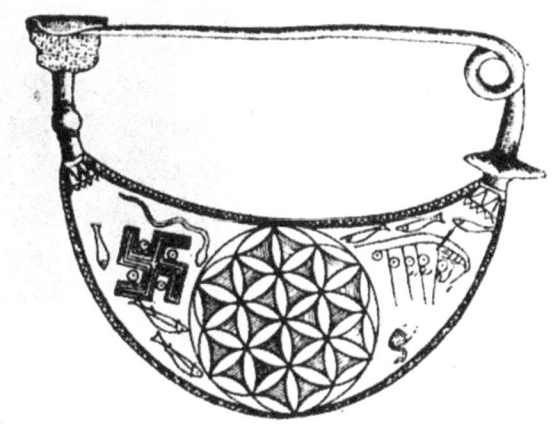

Fig. 8.

Fig. 9.

---

[13] Erwin Bielefeld, "Schmuck", *Archaeologia Homerica*, Band 1, Kap. C (Göttingen: Vandenhoeck Ruprecht, 1968), p. C51, fig. 6j.

This must be compared to two others found in Thebes in Boeotia, dating from the same time and reproduced by Oscar Montélius (fig. 9).[14] In one point, this represents a real progress. The figure shows the circle divided not only in six but in twelve equal angles.

These two questions deserve to be examined in greater depth.

Translated from the French by Kim Williams

---

[14] Oscar Montelius, *op. cit.*, 1ère partie, pl. 24.

Arab and Pisan Mathematics in the Piazza
dei Miracoli[1]

Introduction

Some years ago I began to study the symmetries of the exterior of the Baptistery of Pisa in the Piazza dei Miracoli systematically (figs. 1, 2). The reason for my interest was the evident clash between the symmetries of the interior and exterior of the building, a clash that noticeably reduces its harmony, and which led me to ask myself how this could have come about. The results of this examination were published, thanks to Professor Salvatore Settis, in the *Annali della Scuola Normale Superiore di Pisa*, where the interested reader can find more details.[2]

The aim I set for myself in that study was different than that of this present essay, although this one is based on the article in the *Annali*. The first time my research was centered on the Baptistery, or to be more precise, on its various symmetries and the conclusions that could be drawn from them. The first part of that paper contained an analysis of the symmetries that were visible to the spectator observing the Baptistery from outside. Above all, the observer notes four levels including the dome. As I will explain below, the first and the third levels have a twenty-fold symmetry due to the twenty equally spaced windows; instead, the second level has only twelve windows, and finally the dome with its twelve ribs repeats the twelve-fold symmetry.

The effect of these changes in symmetry is a lack of harmony, a lack of which the architect of the third level was certainly aware: in fact, he surrounded the second level with a colonnade of sixty arches. This sixty-fold symmetry is compatible with both the twelve-fold symmetry of the third level and the twenty-fold symmetry above and below it, because $60 = 30 \times 2 = 5 \times 12$. This analysis of the various symmetries is totally obvious. It is important to bear in mind that the construction of the Baptistery was begun by Diotisalvi on the interior with the erection of eight enormous monolithic columns and four piers according to a twelve-fold symmetry. This means that the builders of the exterior changed the design no fewer than four times, returning twice to Diotisalvi's original design. I must add that there is also a contradiction between the construction of the interior and that of the exterior, a contradiction to which we will return in a moment.

At the time I wrote the paper published in the *Annali*, I asked myself if these changes in design might possibly be used to establish a chronology for the building. A possible chronology was suggested by a thesis by Christine Smith,[3] which I used as a guide for problems of history and archaeology and to whom I express my thanks.

---

[1] Originally published as "Mathematica araba e pisana nella Piazza die Miracoli," pp. 85–99 in *Fibonacci tra arte e scienza*, Luigi A. Radicati di Brozolo, ed. (Milan: Silvana Editoriale).

[2] David Speiser, "Symmetries of the Battistero and the Torre Pendente in Pisa," *Annali della Scuola Normale Superiore di Pisa*, Classe di Lettere e Filosofia Serie III Vol. XXIV, 2–3, 1994: 511–564, with 7 plates.

[3] Christine H. Smith, *The Baptistery of Pisa* (New York: Garland Publishing, 1978).

Fig. 1. The Baptistery of Pisa in the Piazza dei Miracoli, viewed from the Cathedral.
Photo by Carlo Cantini

Naturally my proposal was one, perhaps the simplest, among many possibilities; the final word must come from the archaeologists.

All of the details of this second part of my examinations lie outside the scope of this present paper, and here I limit myself to referring the reader to the earlier paper; as far as the first part of that paper, that is, the analysis itself, is concerned, here I will only present as much as is relevant for what follows.

In this present discussion I will limit myself to drawing the reader's attention to the most interesting mystery that is seen in the Piazza dei Miracoli, that is, what led the builder of the lower level of the exterior of the Baptistery to change Diotisalvi's original design so radically through the choice of a twenty-fold symmetry rather than choosing to carry out the initial design. This first change was decisive for the entire later history of the construction, being as it was in such violent contrast with the interior, which remains consistent with the twelve-fold design of Diotisalvi. For example, as I will show below, the position of the windows of the second level, arranged according to the twenty-fold symmetry, interferes on the interior in an unfortunate way with the original design. We are thus led to ask how the builders could allow themselves to commit an act that was almost sacrilegious in the eyes of the Pisans.

The first answer that comes to mind is that the builder was fascinated by the very unusual symmetry of another important building. It takes no leap of imagination to understand that the building I mean is the Tower of Pisa, and in order to make the novelty of its plan evident I must go back a moment to illustrate the peculiarity of this aspect, which is rarely mentioned in the literature. It is precisely here that the Arab and Pisan mathematicians who were predecessors and contemporaries of Fibonacci intervene.

All of us were taught in high school that while it is quite simple to construct an equilateral triangle or square with a straightedge and compass, as it is to bisect their angles and thus construct hexagons, octagons and so forth, it is much more difficult indeed to construct a regular pentagon (five angles) or an icosahedron (twenty angles), not to mention a pentadecagon (fifteen angles). In order to construct these, one must know the "golden section," which at that time in Europe was probably known only in Pisa, and perhaps in Spain. The golden section divides a segment in such a way that the ratio between the smaller part and the larger part is equal to the ratio between the larger part and the whole segment; by making these two ratios equal what is obtained is a quadratic equation whose solution contains the square root of five, a number that is significant for the construction of the regular pentagon.

Fig. 2. Photo by Carlo Cantini

In order not to become sidetracked from my commentary on the architectural history of the Baptistery, I have included the mathematical details of the construction of the pentagon in Appendix 2, in which I show how, by means of the pentagon and the triangle, it is possible to construct, in addition to the decagon and the icosahedron, the pentadecagon: these three figures play a central role in what I call the drama of the Baptistery, which I will now present to the reader.

The backdrop for the drama is the Cathedral, which establishes the ideal model for the other buildings in the piazza and testifies to the importance that the Pisans attributed to geometric symmetries, and to the mastery with which they used them in their buildings. The drama – that is, the clash between diverse symmetries – is played out in the Baptistery begun by Diotisalvi, the design of which today can only be seen in the building's interior, including, perhaps, the cone which is now covered by the dome. The tower – its pentadecagonal plan inspired by a discovery made by the Greeks and known in Europe to only a few besides the Pisan mathematicians, who in their turn probably learned of it from their Arabic colleagues, and begun when the construction of the interior of the Baptistery was well underway – represents an intellectual challenge to Diotisalvi's design.

The design of the tower was enthusiastically received by Guidolotto, the builder of the first, or ground, level of the Baptistery, who then tried to adapt the idea suggested by the tower to as great an extent as he was able. Because this ground floor clashes violently with the interior, it had to have been met with much criticism, since Guidolotto's immediate successor returned to Diotisalvi's original concept. This dispute between architects continued with two more changes that led to the bizarre building that we see today.

The backdrop for the drama: the Cathedral

Among the three extraordinary buildings of the Piazza dei Miracoli, the Cathedral is without a doubt that which most impresses the visitor. In his *Cicerone* Jacob Burckhardt acknowledged its extraordinary importance with these words:

> The great merit for having been the first to breathe new life into basilica construction, as far as Italy is concerned, is undeniably due to the Tuscans. The sensitivity that distinguished this people in the Middle Ages, and which we may forgive for occasionally resembling the mentality of the builders of the Tower of Babel, was not long in requiring something more than narrow churches whose exteriors were inconspicuous and interiors were preciously decorated; it took its cue from the dignified and monumental. ... For the complete formation of the type, however, a mere bishop's seat was not sufficient; it required all the municipal pride of a merchant republic at the center of world commerce of those times. What Venice was north of the Apennines, Pisa was to the south. ... Furthermore art made one of its great steps forward here. For the first time since Roman times it sought to organize the exterior so that it was at once lively and in harmony with the interior; it steps the facade beautifully and carefully and organizes the pilasters and arches, the upper portions of the transparent galleries, at first wider, then narrowing in accordance with the nave and gable.

There is also an awareness that its pilasters now belong to a new organism … .[4]

The Cathedral established a model of perfection for the construction of later buildings. Here we will limit ourselves to a single observation regarding the facade, the design of which becomes apparent if we imagine it to be inserted into a square and laid out by means of a simple geometric construction (fig. 3).

In the initial square two diagonals are drawn: the horizontal line through their point of intersection divides the square into two equal parts. The diagonals of the upper rectangle are drawn, and a new horizontal line drawn through their point of intersection, which similarly divides the rectangle into two equal parts. Finally, in the upper rectangle are drawn two lines AC and CG.

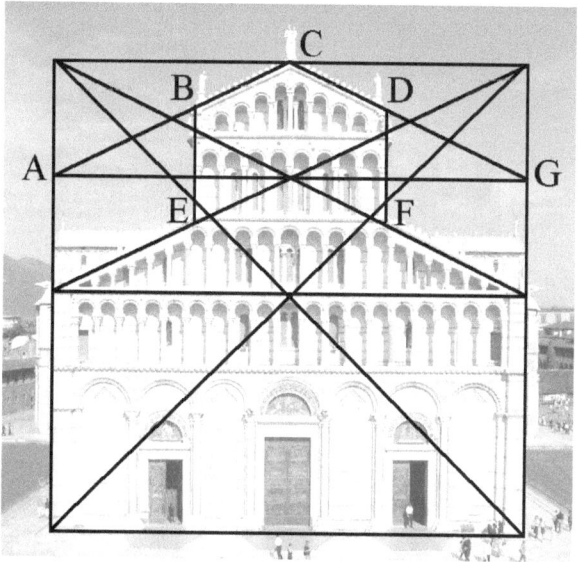

Fig. 3. Hypothesis of the geometric genesis of the facade of the Cathedral of Pisa. It is possible that the positions of vertical lines BE and DF are determined by the golden section, that is, that they are positioned such that AC/AB=AB/BC.
Diagram by Kim Williams

---

[4] *Das grosse Verdienst, dem Basilikenbau zuerst wieder ein neues Leben eingehaucht zu haben, gebührt, was Italien betrifft, unstreitig den Toscanern. Der hohe Sinn, der dieses Volk im Mittelalter auszeichnet, un dem man auch ein stellenweises Umschlagen in die Sinnesart der Erbauer des Thurmes von Babel verzeihen mag, begnügte sich schon früher nicht mehr mit engen, von aussen unscheinbaren und innen kostbar verzierten Kirchen; er nahm eine Richtung auf das Würdige und Monumentale. … Zur vollen Ausbildung des Typus reicher aber ein blosser Bishofsitz nicht aus; es bedürfte dazu des ganzen municipalen Stolzes einer Reichen im Centrum des damaligen Weltverkehrs gelegenen Handelsrepublik. Wie nördlich vom Appenin Venedig, so vertrat südlich Pisa diese Stelle. … Ausserdem aber thut di Kunst hier einen ihrer ganz grossen Schritte. Zum erstenmal wieder seit der römischen Zeit sucht sie den Aussenbau lebendig und zugleich mit dem innern harmonisch zu gliedern; sie stuft die Fassade schön und sorglich ab und giebt dem Erdgeschoss Wandsäulen und Wandbogen, den obern Theilen durchsichtige Galerien, zunächst längere, dann den Mittelschiff und dem Giebel entsprechend kürzere. Sie weiss auch, dass ihre Wandsäulen jetzt einem neuen Organismus angehören ….* (Jacob Burckhardt, *Der Cicerone: eine Anleitung zum Genuss der Kunstwerke Italiens*, Basel: Schweighauserische Verlagsbuchhandlung, 1860, p. 99–101).

Fig. 4. The church of Santo Sepolcro, Pisa. Photo by Carlo Cantini

Fig. 5. The Chapel of Sant'Agata, adjacent to the church of San Paolo,
Ripa d'Arno, Pisa. Photo by Carlo Cantini

That this construction is not a mere curiosity noted post factum but instead represents the basis for the construction of the Cathedral's facade is testified to by the exactness of the lines that are seen in fig. 3. Apart from the two vertical lines BE and DF, the important elements are not the numbers (excluding the number 2 originating from successive divisions), but rather the purely geometric nature of the construction. Instead, the position of the vertical lines poses a problem: it is possible that their position is determined by the golden section, but proving this would require an accurate analysis of either the original design or a precise survey of the actual structure. What is important here is understanding the significance that the architects attributed to the geometric symmetries in constructing the Cathedral. Thus, turning our attention to the other two buildings, we are not surprised to discover mathematically complex symmetries that play an even more important role in the architectural concept. It should be underlined that in these cases as well the significance of the symmetry does not lie in the numbers, but in the division of space.

The scene of the drama: the interior of the Baptistery

As I have already noted, we must begin in the interior of the Baptistery. The enormous dimensions of the eight columns that support the central nucleus indicate that they were erected before the construction of the exterior walls. Only the interior of the Baptistery shows us the original concept of Diotisalvi, to whom Pisa was already indebted for two octagonal churches, Santo Sepolcro (fig. 4) and Sant'Agata (fig. 5). However, in the Baptistery he took a step forward, daring to construct a dodecagonal structure, that is, a centrally-planned building with a twelve-fold symmetry, which represented a new constructive challenge.

The visitor who enters the Baptistery is immediately struck by the upward thrust due to the twelve pilasters and columns that surround the interior space covered by the dome (figs. 6 and 7). Above them is a gallery that in its turn supports twelve piers, and finally above that appears the dome with its twelve openings introducing the decisive element of light from above, in the style that Brunelleschi would later bring to perfection in the Old Sacristy of San Lorenzo and in Santo Spirito in Florence. In this way the interior space is entirely organized according to twelve-fold symmetry, even if the four piers demarcate a square structure.

With respect to the square defined by the four portals, that of the piers is rotated by 45°, such that the portals correspond to an imaginary line that bisects the angles between the piers (fig. 8). The baptismal font, added somewhat later, exploits this structure in a masterful way: four of its sides correspond to the portals, and four to the piers, so that no clash is ever noticed between the octagonal and the dodecagonal structures, except in the pavement, which accentuates the octagonal symmetry of the font. Except for this, all is subordinated to a single, dominant idea and the hexagonal pulpit by Nicola Pisano repeats the same motif.

In the meantime, in contrast to the church of Santo Sepolcro, where the arches are pointed, those of the Baptistery are circular. While these are part of the formal language of the Romanesque style, the powerful upward thrust expresses the new Gothic ideal. This is even more remarkable because in a centrally-planned building the vertical thrust is not contrasted by the other force, that is, the horizontal thrust towards the presbytery that is inherent in buildings with a rectangular plan. It is true that in Santo Sepolcro the use of pointed arches and the austerity of the construction express the Gothic ideal in a more perfect way, but the spectator is nevertheless impressed by the coherence and grace of composition in the Baptistery.

Fig. 6. Interior of the Baptistery of Pisa. Photo by Carlo Cantini

However, some dissonances can be noted in the harmony of the central nucleus. The first thing that the observer cannot fail to notice is the position of the windows, which do not correspond to the system of columns, piers, and arches. The dissonant effect produced by the windows of the first level does not perturb the harmony to a large degree, but the clash is accentuated on the second level, and this may perhaps be the reason why some of them were walled up. A greater surprise awaits the visitor who goes up to the gallery, where he discovers windows that cannot be seen from the ground floor because they are concealed by the floor of the gallery. These twelve windows exactly correspond to the columns and piers. If then the visitor is lucky enough to be allowed inside the dome, he will discover that the Baptistery has two coverings: under the dome lies a cone that, here as in Santo Sepolcro, could serve as a roof. To complicate matters even further, there seem to be traces of two foundations. Fortunately, we do not have to deal with these complications here.

Fig. 7. Section of the interior of the Baptistery of Pisa, from *Le Fabbriche Principali di Pisa ed alcune vedute della stessa città, 24 tavole* by Ranieri Grassi (Pisa: Edizioni Ranieri Prosperi, 1831)

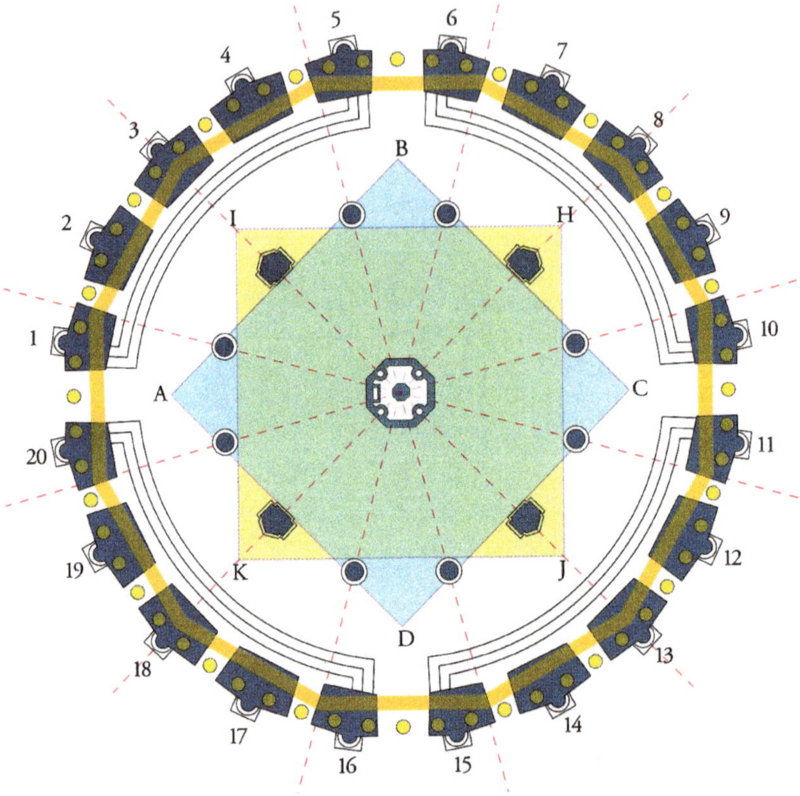

Fig. 8. Baptistery: Plan with geometric construction. With respect to square ABCD defined by the four piers, square IHJK, defined by eight columns, is rotated 45°. The intersection of the two squares forms an octagon similar to that of the baptismal font. Neither the dodecagonal symmetry (four piers and eight columns) nor the octagonal symmetry is compatible with the twenty-fold symmetry formed by the twenty sides of the exterior perimeter, as shown by the red radii that intersect the columns and piers. In contrast, the twelve radii are compatible with the vertices of the perimeter (shown in yellow) of the dodecagon of the second level. The sixty columns (in yellow) of the arcade of the second level are compatible with both the twelve sides of the second level (five per side) and the twenty of the first level (three per side). Diagram by Paolo Radicati di Brozolo

We begin to suspect that the organization of the exterior must differ from that of the interior: on the exterior there must be one additional level, and its symmetry must be different. If now the visitor, surprised and amazed, exits the building in order to verify his suspicions, he will not be disappointed. The double covering explains the bizarre silhouette of the dome: three levels than be discerned on the exterior in correspondence to two on the interior. Even more important, however, and even stranger because they are in violent contrast to the interior, the first and the third levels show a twenty-fold symmetry rather than twelve-fold.

The challenge of the Tower

The lower level of the Baptistery was intentionally built as a challenge to the pre-existing structure of the interior: its four piers and eight columns had clearly established a twelve-fold symmetry, and perhaps the construction of the interior was already well on its way. Thus we are dealing with the most surprising of all the violations of symmetry in the history of the Baptistery, and the one that is also the most interesting for at least two reasons. First of all, whoever wanted to challenge Diotisalvi's design did not have any particular motive or necessity to do so: to the contrary, the concept of the interior was beautiful, and we have to believe that the design for the building's exterior was equally as elegant, and that Diotisalvi's successors had every reason to carry out the design of their master. Thus it follows that there had to be an important reason to induce them to such an alteration. Further, while the second and third levels are of inferior quality, the exterior of the first level is entirely comparable in both its design and its execution, to that of the interior, but the aesthetic aim is quite different. The strong vertical tendency of the interior gives way on the exterior to an organization of the circular wall that masterfully counterbalances the circular and the vertical directrixes. Thus we are led to suspect the presence of a leading artistic personality, one so audacious that he dared to challenge and alter the earlier design, replacing it with one of his own. We must therefore look for an event of the greatest importance, one capable of providing the architect and his supporters with the inspiration for such an alteration. Now, what was the most important artistic event of the times, one which certainly impressed the Pisans, and the artists in particular? To find the answer we need not look far: behind the Cathedral, still today one of the greatest and most splendid, shortly after the foundations of the Baptistery were laid, the Pisans had begun to construct what they intended to be the highest tower in the world (fig. 9).

But before explaining what might have been the motive that drove the builders of the exterior of the Baptistery to follow the example of the Tower, we have to examine the architecture in greater detail. The style of the Tower is Romanesque, and its constructive and decorative elements – in particular, the round arches – are drawn from the Romanesque vocabulary. However, the artistic concept differs from other Romanesque towers, which generally become lighter, more open, and more transparent as they rise in height. For example, often only a small opening is found at the lowest level, then two, then three and so on as the tower rises, until towards the top the openings are both more numerous and larger in size. This does not happen in the Tower of Pisa: except for the doubling of the arches at the second level and the fact that the columns at the seventh level are slightly taller, all of the levels are rigorously equal. This translational symmetry produces a strong unifying effect (unfortunately, the fact that the Tower leans reduces this effect to a great extent). Translational symmetry was used as an aesthetic element in all cathedrals, but only in a horizontal direction. The Tower of Pisa is thus an example of a building where a new idea – translational symmetry – was used in a vertical direction rather than in horizontal repetitions, and was expressed by means of an earlier formal device – the Romanesque arch – already present in the Cathedral.

In a recent book Piero Pierotti[5] has shown the importance that the Pisans attributed to measurements that were numerically exact: the height of the Tower is exactly 100 Pisan *braccia*, equivalent to 20 Pisan *pertiche*, while the circumference measures exactly 100 Pisan *piedi*. These numbers are proof of how much value the Pisans of that period placed on precise mathematical relationships.

---

[5] Piero Pierotti, *Una torre da non salvare* (Ospedaletto, Pisa: Pacini Editore, 1990).

Fig. 9. The Leaning Tower of Pisa. Photograph by Carlo Cantini

There is another interesting division of space in the Tower: according to Pierotti, the heights of the eight levels are respectively 5, 2, 2, 2, 2, 2, 2, 5 *pertiche*. There is thus, in addition to the translational symmetry, a mirror symmetry that accentuates the close-knit consistency that this partitioning confers on the Tower. The Tower thus presents itself as a unitary entity, albeit one whose magnificent solemnity is unfortunately reduced by its leaning.

In contrast to other bell towers, the Tower of Pisa's perfect translational symmetry in the vertical direction inevitably induces the spectator to count the number of floors, of which, including the bell compartment, there are eight. Why precisely eight? Many reasons have been suggested, some based on theology, but none of them seem convincing to me. We have to remember that when the construction of the Tower was begun, it was intended to be the highest building in Italy, and perhaps in Europe, and as such, represented the immense "pride" of the Pisans of the period.

Now, what was the highest tower that the Western world had known? Naturally, the Tower of Babel, the example *non plus ultra* of a very high building. The Bible does not say much about the Tower of Babel's construction, but says a lot about its destruction. Herodotus, always rich in information, explicitly says that it had eight levels, and in the first book of his *Histories* writes: "In the middle of the sanctuary has been built a solid tower, a stade long and the same in width, which supports another, and so on: there are eight towers in all."[6] There is a missing link to prove that the Pisans truly intended to imitate the Tower of Babel, and that is, there is no proof that they knew of Herodotus's description, since his book was not known in its entirety to either the Western world or to the Arabs. On the other hand, it cannot be excluded that some parts of it were known in the West, perhaps via Constantinople or Salerno. It even seems, as Professor Giuseppe Nenci told me, that a manuscript of Herodotus's work existed in a monastery not far from Pisa.

While a consideration of translational symmetry took us all the way to the Tower of Babel, an examination of the rotational symmetry leads us to another great – and for us, more important – surprise. The lower floor of the Tower consists of a massive wall to the exterior of which is applied a blind arcade. Above this there follow six more levels, each equal to the others, except for the last, which is slightly taller. Each of these levels has, with respect to the lowest, double the number of arches which circle the exterior wall like a colonnade; at the top is a much smaller tambour. There are thus eight levels in all. The finite symmetry of the Tower is naturally determined by the number of pilasters and arches.

To our great surprise, we find that the number of arches is not sixteen, as we might have expected, nor twelve or even twenty-four (which represent the most common and easily achieved symmetries). There are instead fifteen arches at the lowest level, and thirty on the others! Finally, the tambour has six large and six small arches arranged such that the smaller embrace one arch, and the larger embrace four of the seventh level. This is an extremely rare, if not unique, example of fifteen-fold and thirty-fold symmetry; I know of no other towers with fifteen sides, nor of any other example of fifteen-fold symmetry. Such a rarity cries out for an explanation, even more so precisely because these symmetries are used in such an extraordinary building. Since this fact, as far as I know, has never been pointed out, except by Piero Pierotti in the book mentioned earlier, it comes as no surprise that no explanation has been proffered. Peculiar elements in religious buildings are often, though not always,

---

[6] Herodotus, *The Histories*, Robin Waterfield, trans. (Oxford: Oxford University Press, 1998), I, §181, p. 79.

explained by theology. This kind of explanation does not, however, seem valid in this particular case. There are four major prophets, four evangelists, four Fathers of the Church, seven archangels, twelve tribes, twelve apostles; on the other hand there are many more than fifteen popes, an even larger number of saints, and still more angels. Instead, fifteen has never been a theological number, and thus we must look elsewhere for an explanation of the truly singular choice.

It is at this point that we must think of the Arab and Pisan mathematicians who lived in Pisa at and around the period of Fibonacci, mathematicians who most certainly knew how to construct a regular pentagon and a pentadecagon, and who were well aware of being probably the only ones who possessed this knowledge. Thus, in my opinion, the explanation for the artistic concept of the Tower, as well as for the abandonment of Diotisalvi's design for the construction of the exterior of the Baptistery must be sought in the very close collaboration between artists and scientists in Pisa of that day. To be sure, such an idea must have been pleasing to the proud Pisans.

The name of the architect of the exterior of the Baptistery appears to be unknown: if it were Guidolotto himself, his prestige would explain how he was able to make the abandonment of the original design acceptable. Finally, we must ask ourselves how the architect of the Tower was able to construct the pentadecagon in practical terms. There are at least two possibilities: the simplest consists in using a cord AB that is as long as the intended perimeter of the Tower (fig. 10). On a second cord of arbitrary length that begins at A, fifteen equidistant points are marked; let C be the fifteenth. Now consider the triangle ACB, and from all the points of the second cord draw the parallels to side CB. The intersections of these lines with AB form a series of fifteen equidistant points, and thus we have obtained the division of AB into fifteen equal parts.

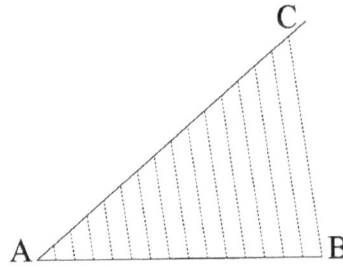

Fig. 10. A simple method for dividing a given segment AB into fifteen equal parts.
Diagram by Paolo Radicati di Brozolo

But flanking this purely technical solution there is another, consisting in carefully leveling the terrain and then constructing on it the polygon in the manner that will be explained in Appendix 2. This method would have offered an additional advantage which I think may have come into play. Effectively, it is easy to draw about a center – that is, exactly beneath the Tower's axis of symmetry – a pentagram, or five-pointed star, which explicitly shows the golden section. This pentagram is found on the facades of many cathedrals and churches, especially north of the Alps, but also on the Cathedral of Milan. What is the significance of this symbol? The answer is, as readers of Goethe's *Faust* know, that the pentagram has the power to keep the devil away. Who knows! In any case, the men who used to watch the opening festivities of the annual fair from the galleries that gird the Tower, dressed in their most elegant robes, would certainly have felt more secure!

What happened next: the duel between two artistic conceptions fought on the exterior of the Baptistery

In 1184 Diotisalvi's design for the exterior of the Baptistery was replaced by that of Guidolotto, who, even though not having himself designed the Tower, was nevertheless strongly influenced by it. His design was conceived with a radically new idea based on Pisan knowledge of geometry, that is, on the construction of a regular pentagon and pentadecagon, used in the construction of the Tower. This concept remains unique in the history of European architecture.

The first level of the Baptistery's exterior is the immediate result of the drama, leaving scars on the building which are still painful to see today. However, it would be unjust not to acknowledge the harmonious arrangement of the twenty arches, in spite of the fact that the openings are so very narrow. Here it may be that the Romanesque architecture comes closer to that of the Renaissance than in other monuments. This is due precisely to the choice of the proportions, and it would be worthwhile undertaking an accurate survey to try to discover Fibonacci ratios or perhaps golden sections. It may be that it was in order to achieve these proportions that the height of this level is less than that of the interior.

It can be hypothesized that Giudolotto's design for the exterior followed that of the Tower, and included two ulterior levels of equal or almost equal height, each having twenty or perhaps forty arches. This would have been crowned with a dome that was perfectly circular such that from the outside the spectator would not have been able to perceive any conflict between the twenty-fold symmetry of the exterior and the twelve-fold symmetry of the interior. The first level that we see today leads us to think that such a design would have been truly very beautiful.

However, Guidolotto's "*coup d'etat,*" while probably enthusiastically accepted by many, must not have been pleasing to everyone. In 1210 there were heated arguments that led, perhaps together with other reasons, to battles between the archbishop and the clergy that lasted for about ten years. When the dispute between the archbishop and the Capitolato ended around 1221, there was a return to Diotisalvi's original idea, if not to his precise design, and thus the twelve-fold symmetry triumphed. It is possible that the colonnade, with its sixty-fold symmetry, was added at precisely that time, even if, as I mentioned earlier, it is more reasonable to posit a later date. However, the problem is very closely connected to that of the lack of correspondence between the number of levels on the interior (two) and the exterior (three). The answer can only be found through archaeological research.

A third change in the design was made around or just prior to 1278. The twenty-fold symmetry won again, but the quality of the execution of the third level was decidedly inferior to that of the first. Around that time, or just after, the thirty skylights decorated with sculpture by Giovanni Pisano were added, ruining, however, yet another time, the symmetry. As can be seen in fig. 11, to each of the twenty arches of the first level there correspond three of the sixty arches of the colonnade, but the symmetry of the thirty wimpergs is no longer compatible with the twenty arches of the first level.

How and when the two coverings, the cone and the dome, were added, is left to the archaeologists to determine. At that time the twenty-fold symmetry was excluded by necessities of the construction, but it would have been possible to construct an external covering such as that of San Paolo in Ripa d'Arno (see fig. 5), and which was probably the way that Guidolotto had designed it.

Fig. 11. Exterior view of the second-level gallery on the exterior of the Baptistery of Pisa. Photo by Carlo Cantini

If we contemplate the Baptistery today, the loftiest inspirations and the most convincing results seem to be those of Diotisalvi's interior and Guidolotto's lowest level on the exterior. Each in itself represents an extraordinary and admirable conception, but unfortunately they contradict each other in a way that we would almost want to call fratricidal. In consequence of this contrast there was a series of returns to previous conceptions and compromises in search of a reconciliation. Still later were added stupendous masterpieces such as the baptismal font by Guido Bigarelli and the pulpit by Nicola Pisano, giving the interior an impression of true unity.

As far as the exterior is concerned, we might say that the level of artistic quality deteriorated as it progressed because the first level is the most perfect. Even so, in moonlight the Baptistery makes a truly extraordinary impression on the visitor.

Final observations

At this point the reader who has followed the exposition of this drama attentively will probably ask himself, "All of this appears interesting, but what parts of it are true? Are there perhaps not too many hypotheses that are unconfirmed by written documents?" I admit, my reconstruction of events is based on hypotheses which are not backed up by documents, but which are documented by the monuments themselves.

To be sure, the three clashes between the twenty-fold and the twelve-fold symmetries on the exterior are right in front of our eyes. The same is true for the fifteen-fold symmetry of the plan of the Tower. That the architect learned the construction of the pentagon and the pentadecagon from Pisan mathematicians, and perhaps directly from the Arabs, this too is beyond doubt. Everyone can judge for himself whether or not the aesthetic of Guidolotto differs from that of Diotisalvi, but in this case as well it is beyond a doubt that the two architects' methods for expressing these ideas were different. However, even though I am not able to provide documentary evidence, it is difficult to imagine that the Pisans were not influenced by the design of the Tower. These are essentially the only two hypotheses on which I base my exposition.

Appendix I

So that the reader will not think that I based the discussion of the clash of symmetries on some "Meistersingers' rule" deduced from an orthodoxy that was invented ad hoc, here I want to cite two examples where different symmetries and partitionings are combined in harmonious ways.

Fig. 12. The refined combination of different symmetries in the dome of San Lorenzo in Turin. Photograph by Kim Williams

The first (which perhaps might have influenced the construction of the Baptistery) is the so-called Mosque of Omar (more precisely, the Mosque of Abd al Malid) built in Jerusalem in 687. In that extraordinary edifice, the sixteen-fold symmetry of the sixteen columns of the interior colonnade is surrounded by a second colonnade of twenty-four elements (twenty-four-fold symmetry). However, the stupendous harmony of the building is not disturbed, because of the extraordinary play of squares, octagons and circles.

The second example is that of San Lorenzo in Turin (fig. 12), which is perhaps the building in which the most refined combination of diverse symmetries is found. Guarini surrounded an octagon with star pentagons and star hexagons so that not only do the different symmetries not generate any dissonances, but to the contrary, they produce an effect that is surprisingly harmonious. It is therefore one thing to combine different symmetries harmoniously, and another to look for a clash between them.

Appendix 2

The construction of regular polygons has always attracted the interest of artists and mathematicians; indeed, we can say that for a long time it was the most important problem in mathematics. The exact construction of the square probably dates back to the most ancient civilizations; that of the hexagon is well documented in the Mycenaean culture, thanks to a series of medallions dating from the sixteenth century B.C.;[7] finally, the Greeks knew the construction of the pentagon. In what follows, by "pentagon" we always refer to the regular pentagon; it is worthwhile to underline that by "construction" we always mean a construction with straightedge and compass.

Since it is simple to bisect an angle, when we know how to construct an m-gon, we know how to construct a $2m$-gon. For example, when we know how to construct a pentagon, where $m = 5$, then we know how to construct a decagon, where $2m = 10$. This kind of construction does not exist, for example, for the 7-gon (heptagon) and the 9-gon (nonagon).

The construction of the pentagon is relatively complicated, and we will describe it starting from a special property of the pentagon, or better, of the pentagram, that is, the star inscribed in the pentagon.

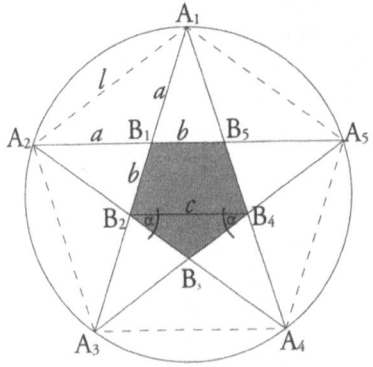

Fig. 13. Diagram by Paolo Radicati di Brozolo

Let l be the side of the pentagon: by joining each vertex A_1, A_2, A_3, A_4, A_5 with the two vertices opposite it we obtain the pentagram. It is not difficult to prove that $A_1B_1 + B_1B_2 = A_1A_2 = l$; in like manner, given a as the length of A_1B_1, we can prove that $a = A_1B_1 = B_2B_4$. It thus follows from the similarity of the two triangles $A_1B_2B_4$ and $A_1B_1B_5$ that $l : a = a : (l - a)$ (fig. 13).

Point B_1 divides segment A_1B_2 of length l in two parts a and $b = (l-a)$ in such a way that the whole (l) is to the larger part (a) as the larger part (a) is to the smaller part (b). This is the famous golden section.

When l is known, the ratio $l : a = a : (l - a)$ (a second-degree equation) makes it possible to calculate a. The solution is:

$$a = l/2\left(-1 + \sqrt{5}\right) \cong 0.618\,l .$$

Now that this property of the pentagon has been established, we propose to construct one side l. From end E of a segment EF of length l we construct a circle tangent to EF with radius $l/2$ and center O. By the Pythagorean theorem, the length of segment OF is

$$\sqrt{l^2 + \frac{l^2}{4}} = l\frac{\sqrt{5}}{2} .$$

Segment EF is thus divided into two parts: FK, of length a, and KE of length $l - a = b$ (fig. 14).

---

[7] See "The Symmetry of the Ornament on a Jewel of the Treasure of Mycenae" in this present volume, pp. 1–8.

At this point the construction of the pentagon is easy. Going back to fig. 13, from point A_2 is drawn a segment A_2B_5 of length $l = a + b$ divided, with the method described above, in two parts A_2B_1 and B_1B_5 of lengths a and b respectively.

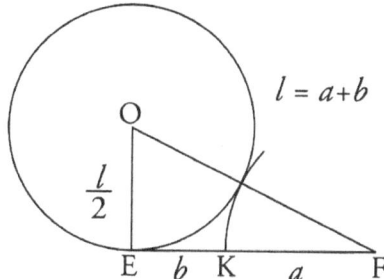

Fig. 14. Diagram by Paolo Radicati di Brozolo

Segment A_2B_5 is then prolonged with a segment of length a to point A_5. Point A_1 is then determined as the intersection of two circles with centers in B_1 and B_5 and radius a. Then a line is drawn through A_1 and B_1 and point B_2 is marked at distance b from B_1. The pentagon is then completed.

The pentadecagon

To construct a pentadecagon we begin with the figure shown in fig. 13. The center of the circumscribing circle is the intersection O of the two bisectors of the angles at A_1A_2. Next is constructed equilateral triangle A_1HK, whose interior angles are 60° (fig. 15).

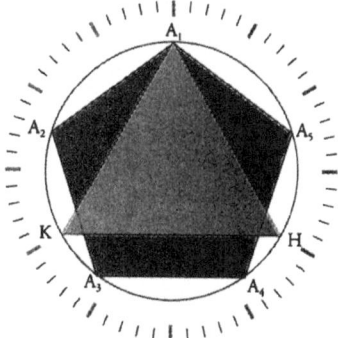

Fig. 15. The construction of the pentadecagon from a pentagon and equilateral triangle inscribed within the same circle.
Diagram by Paolo Radicati di Brozolo

Let us now imagine that the circle circumscribing the pentagon and the triangle is the face of a clock. The vertices of the triangle correspond to 0′, 20′ and 40′, while those of the pentagon correspond to 0′, 12′, 24′, 36′ and 48′. Arc HA_4 thus corresponds to 4′, or a fifteenth of the entire circumference, and the segment HA_4 is thus the side of the pentadecagon.

Translated from the Italian by Kim Williams

Introduction

The subject I am going to talk about here belongs, one may say, to the prehistory of descriptive geometry: it is part of our modem discovery of space. Three times a civilization has made such an investigation: in ancient Egypt, in Antiquity, and in modern times, where perhaps we should speak of space-time. And each time, not only science, but the arts participated in this endeavor as well. It is always extremely interesting to compare the progress of the sciences with the evolution of the arts, as well as their histories, their results, and their methods. But it is fair to say that in spite of many valiant pioneering efforts, so far this has not been done systematically enough: think for instance of medieval architecture and its importance for the progress of technology and science.

This small contribution is devoted to two mathematical – that is, geometric – discoveries made in 1503 and 1504, and presented in two famous paintings by Raphael: *Lo Sposalizio* (*The Marriage of the Virgin*) and *L'incoronazione della Vergine* (*The Crowning of the Virgin*). It is especially in the second one that we find architecture, mathematics and theology closely intertwined in a way that is deeply characteristic of this artist, who we will also see here as a great scientist.

La dolce prospettiva

Attempts to represent buildings in perspective go back at least to Giotto and his school. But it seems that around 1400 Masaccio was the first to discover the law of the vanishing point; I remind you here simply of the Christ on the cross in the *Holy Trinity* in Santa Maria Novella and his frescoes in Santa Maria del Carmine, both in Florence. North of the Alps, the early Flemish painters approached this law step by step, by trial and error. This process is described in an essay by Erwin Panofsky; it seems that the first correct painting is Dieric Bouts' *Last Supper* in St. Peter's in Leuven. However, in all these paintings we find only the use of what is sometimes called, a bit misleadingly, "central perspective." This means that all buildings are presented to us frontally, and the horizontal edges are either orthogonal to our view, in the line of our view, or converging with it. Thus, there is always only one "vanishing point," the point towards which the parallels converge. A typical example is the *Delivery of the Keys to St. Peter* by Perugino in the Sistine Chapel (fig. 1).

[1] Originally published in *Nexus III: Architecture and Mathematics*, Kim Williams, ed. (Ospedaletto, Pisa: Pacini Editore, 2000), pp. 147–156. The paper was presented at the symposium "Omaggio a Edoardo Benvenuto" organized by Profs. Massimo Corradi, Orietta Pedemonte and Patricia Radelet-de Grave, 29 November–1 December 1999, Genoa, Italy, and at the third international conference "Nexus 2000: Relationships between Architecture and Mathematics," 4–7 June 2000, Ferrara, Italy, directed by Kim Williams.

Fig. 1. Perugino, *Delivery of the Keys to St. Peter*, Sistine Chapel, Vatican City. Reproduced by permission, Musei Vaticani

Fig. 2. Raphael, *The Crowning of the Virgin* (Oddi Altarpiece), Pinacoteca Vaticana, Vatican City. Reproduced by permission, Musei Vaticani

Please note that this restriction forced the painter to place all buildings parallel to each other and frontally with respect to the observer: a severe restriction indeed! So we may ask: who was the first painter who succeeded in correctly representing a building in other than the frontal position?

Perugino's fresco dates from 1480/81, and in a moment you will see a second, very similar one. But in 1503 his pupil, Raphael Sanzio, was invited to paint a *Crowning of the Virgin* for the church of the Franciscans in Perugia, which is today in the Pinacoteca del Vaticano (fig. 2). I think that this is the first painting where a structure in a non-frontal position, the sarcophagus of the Madonna, is constructed rigorously. At least I have never seen an earlier one myself. So the question arises: how did Raphael do it? How did he achieve what so many others, presumably, had tried to do in vain?

But first: can we be sure that the sarcophagus of the Madonna is constructed correctly? It is fairly easy to convince yourself that the long edges do indeed converge to a vanishing point. For the short edges, this is obviously a bit more difficult; I convinced myself that they do so, but it seemed that the vanishing point to the left lies a tiny bit higher. But this may be due to my clumsiness together with the fact that I had to work with a comparatively small reproduction, or it may be due to the fact that, according to Jones and Penny,[2] the painting was transported from wood on linen.

So how did Raphael do it? You can see the answer in fig. 3: draw the crossing of the extended shorter edge at right with the horizontal that passes through the summit at left, and then descend from the upper summit to this horizontal line and extend it beyond.

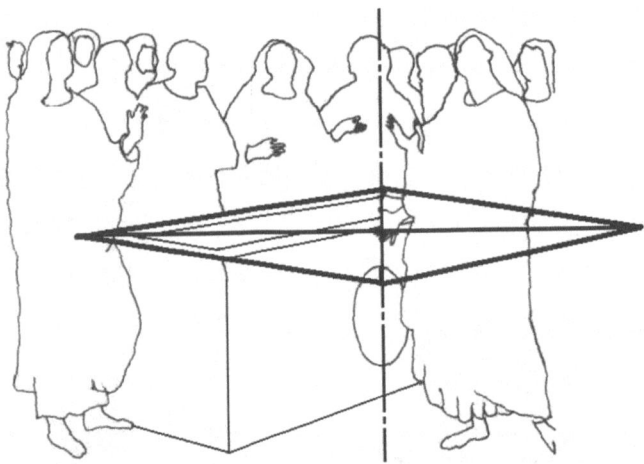

Fig. 3. The perspective scheme of the sarcophagus in *The Crowning of the Virgin*.
Drawing by Kim Williams

Now you see that this extension covers two lines in the painting: one that lies in the horizontal plane, and a second one that ascends vertically from the central crossing point through the center of the right side of the sarcophagus! This means that the central crossing point is the center of the two squares. In the next figure you can now see how Raphael proceeded (fig. 4).

---

[2] Roger Jones and Nicholas Penny, *Raphael* (New Haven and London: Yale University Press, 1983).

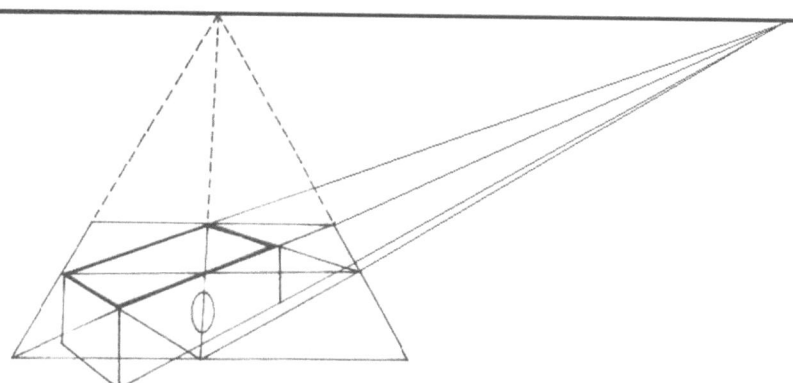

Fig. 4. Perspective construction of the sarcophagus. Drawing by Kim Williams

Raphael first drew a square, then he drew the diagonals that, finally, permitted him to draw the square whose corner points are the centers of the four sides. As often happens, it is all very simple, but one has to think of it! And this construction, while special indeed, is not only the very special case it might seem to be: namely a rectangle composed of two squares. You can, for example, extend the sidelines beyond the half-square and construct any rectangle with a line towards the vanishing point.

Having found the law of the two new vanishing points, Raphael has solved the problem of correctly constructing the right angle of any rectangle that is bisected by a line towards the central vanishing point. My colleague Mario Howald-Haller showed me a trick that allows one to draw correctly a building seen from an arbitrary angle. As he told me, it is contained in the writings of Leon Battista Alberti. That such a solution was looked for, can be seen from paintings that try to give the impression of a building not placed frontally with respect to the spectator, although it is really placed this way. An example is Titian's famous *Madonna with Members of the Pesaro Family* in Venice: there this effect is achieved simply by placing the vanishing point far to the left, outside the painting.

But buildings at which one looks obliquely remained rare for a long time. And in many of them a not-frontally-placed building is painted so that one cannot easily check its construction.

Now: theology. The painting does indeed make a theological statement, but one that is in accordance with the instruction of the Franciscans, who commissioned the painting: the Virgin is placed on the same level as the Christ, not lower, which is unusual. If Raphael himself made a theological statement here, it might be as I am about to explain, but I would not press the point. The axis of the geometric construction does not coincide with the axis of the painting itself. But the former concerns an earthly matter only, while the axis of the painting is determined by the heavenly order. Always remember, especially when we now go to *The Marriage of the Virgin*, that an altarpiece is a symbolic construction, and not only *un coin de la nature vu par un tempérament*! Indeed, at the time, some may have found Raphael's innovation too naturalistic.

Fig. 5. Raphael, *The Marriage of the Virgin*, Pinacoteca di Brera, Milan. Reproduced by permission, Italian Ministery for Cultural Goods and Activities

The Marriage of the Virgin

Lo Sposalizio, or *Marriage of the Virgin*, originally painted for Città di Castello, is now in the Brera in Milan (Fig. 5). The High Priest celebrates the marriage of the elected Joseph with Mary. Only Joseph's stick bursts into flower, and thus his companions break their own sticks, which remained barren. But what catches the eye more than everything else is the building designed by Raphael, its grace, its lightness and, indeed its elegance: one must look far ahead, way into the eighteenth century, to find such a graceful building. Many things contribute to it: the cupola; the coloring; the elegant arcs (the same that Michelangelo would use in an inverted position on the sarcophaga in the Medici Chapel).

But if you look more closely, you see that the lightness is especially due to one accomplishment, with which the pupil outdid his master, Perugino. Rather than the master's slightly heavy octagon, he constructed (and he was the first to accomplish this) a hexadecagon; the building has sixteen sides! And this, as I shall now show, was no mean achievement. It is well known that a regular polygon whose order is a power of two can be constructed simply by bisecting a number of times successively an angle with a ruler. Starting from this result, the construction of such a polygon in central perspective must be obtained in two separate steps. Recall that on the horizontal line you can always assume Euclidean geometry to be valid, and thus I have indicated for the octagon the relevant lengths, which you can transfer directly onto the frontal line. But while this Euclidean construction is almost trivial for the octagon (Figs. 6a and 6b), for the hexadecagon it is more cumbersome: you must construct the equivalent of the extraction of a square root of a term which contains a square root (Figs. 7a and 7b)!

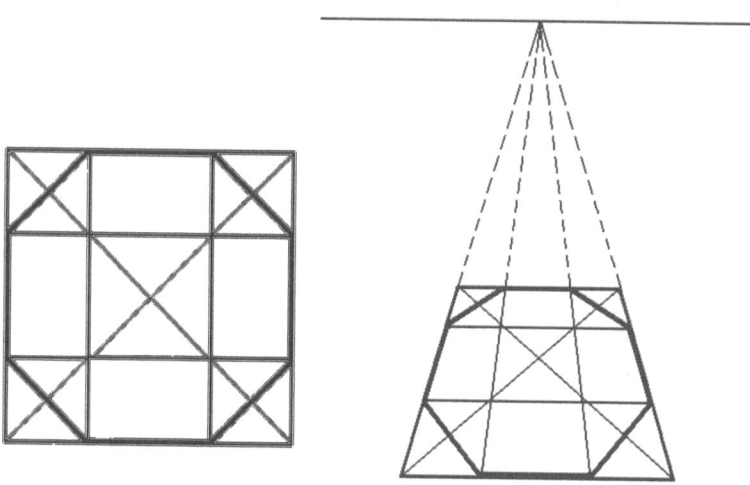

Fig. 6. Perspective construction of an octagon. Drawing by Kim Williams

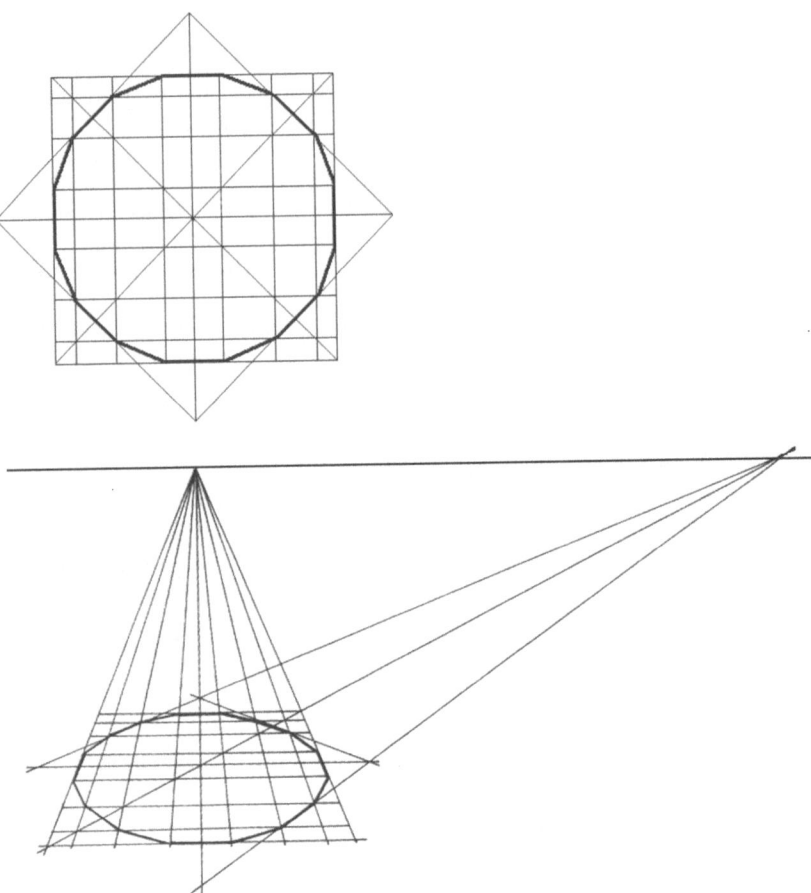

Fig. 7a, b. Perspective construction of a hexadecagon, Drawing by Kim Williams

Then the perspective construction proper must follow; there are two ways to do it. You freely choose the angle from which you see the square (for example, you may choose the rear edge) and with either method you then draw the diagonals. The first way to construct the perspective is to extend the diagonals to the vanishing points found by Raphael, then draw the other, lower edges, then the lines to the central vanishing point, and finally the second horizontal, which yields the last two edges. Mathematically, this is the more transparent procedure, but you need to know the exact location of the new vanishing points, which, as was seen, are often way out of the painting. This difficulty can be avoided by first drawing the lines from all points marked on the frontal edge to the central vanishing point, and then the horizontals that yield the other edges. This method has the advantage that the painter need not work outside the painting. Which way Raphael actual used remains a guess. Indeed, both steps are fairly complicated for this figure, but Raphael worked rigorously: all parallels on any side of the hexadecagon meet at the same vanishing point; this is a grand accomplishment.

Theology in *The Marriage of the Virgin*

The investigation of theological points made in works of art was introduced in this century by Aby Warburg and Erwin Panofsky as part of the broader concept "Iconology."[3] How does theology come into *The Marriage of the Virgin*? I will begin with what you may call the lowest level of it, which is, I think, recognizable in the fact that the heads of the Virgin's five bridesmaids form a regular pentagon, and those of the four companions behind Joseph a square. Of course, this may be due to nothing more than Raphael's well known desire to use geometric figures in his paintings for arranging the people in the painting, especially in depth, as well as to his unrivalled virtuosity of drawing in perspective. But he who draws (or, as here, forms) a pentagon, also draws a pentagram. Thus, where on so many churches, usually on the western front, we find a pentagram, even where it is artistically not especially wanted, there must be a special reason for it. As examples I mention Hanover, Strasbourg, Breisach, Basel and the largest cathedral in Italy at Raphael's time, the Duomo di Milano. No doubt, these pentagrams had a function: they served to keep out the devil, the *demonio*. Thus I suspect that the bridesmaids here are performing a pious service to the Virgin: they protect her; just as the four friends form a tower of strength for Joseph. But again, I will not press the point, and I will pass right on to the higher level, that is, to the real theology, where we stay on safe ground.

Before doing so, I must say a word about the Western tradition of the representation of the Temple. There are two aspects to this tradition: one aspect follows the Biblical descriptions of Solomon's Temple, and Ezekiel's vision. The other aspect, which Raphael follows, goes back mainly to the crusaders and is based on what they saw, namely, the Dome of the Rock, the octagonal mosque by Abd al-Malik, which they imagined to have been built in the tradition of Solomon's Temple, the seat of Wisdom. As such, it became a symbol, not only of the church and of wisdom itself, but also of Mary, the seat of true Wisdom. Still today in Louvain-la-Neuve the academic year is inaugurated *au nom de Notre Dame Siege de La Sagesse*. This tradition also found an expression in numerous altar chapels, tabernacles and so forth, which always stand for the Temple, the Wisdom and the Virgin in one, as you can see in the interesting book by Paul von Naredi-Rainer.[4]

But while the real Temple in Jerusalem was directed towards the East, since Yahweh enters into his house from the East, Raphael orients the front towards the West, as you can see from the shadows cast by the figures; thus the spectator looks towards the East.

I come now to the second step which Raphael took beyond his master, who also painted a picture with an octagonal temple, but the subject of which is a *Sposalizio*, now in Caen (fig. 8). Both paintings were finished in the same year, and there is no doubt that the master had begun his considerably larger painting much earlier. Thus priority in choosing the subject belongs to him, as well as the idea of arranging the couple and the high priest frontally before the temple. In both paintings, contrary to Perugino's fresco in the Sistine Chapel, the door of the Temple is open, and there, in the opening, lies the vanishing point, so that you can follow the lines of view in the painting towards it. Is this innovation with respect to Perugino's first fresco due to the master or to the disciple? I do not know, and the question seems to be a difficult one to decide. In any case it was Raphael who saw the artistic possibilities that this seemingly small step permitted. Perugino's painting is arranged in two frontally

---

[3] Erwin Panofsky, *Studies in Iconology* (New York and Evanston: Harper & Row, 1962).
[4] Paul von Naredi-Rainer, *Salomons Tempel und das Abendland* (Köln: Dumont Verlag, 1994).

oriented layers: the group of the people, more or less on one line before the Temple, and the Temple itself, which serves mainly as a historical indicator for the narration.

Raphael's painting shows, as the German historian Hiller von Gaertringen says, *Tiefensog*,[5] that is, a pull towards the pictoral depth. Besides breaking the one line of people up into geometrically composed groups, he accomplishes this by constructing these many squares, forcing the spectator to follow their edges with his eyes. And these edges meet where? Why, at infinity! And who has his seat at the infinitely distant? Of course: HE, God! Looking towards the vanishing point, you look towards infinity: you look towards God.

Fig. 8. Perugino, *The Marriage of the Virgin*, Musée des Beaux-Arts. Musée des Beaux-Arts de Caen, Martine Seyve photographe. Reproduced by permission

---

[5] Friedrich Freiherr Hiller von Gaertringen, "Raphaels Lernerfahrungen in der Werkstatt Peruginos," *Kunstwiss. Studien* Bd. 76 (Deutscher Kunstverlag, 1999), pp. 56–66.

It is worthwhile to pause here. We often say that parallels meet at infinity, but this is not so according to Euclidean geometry. The figures can extend to the infinite, but Euclidean geometry makes only asymptotical statements about it, and this holds even more for mechanics, which it underlies and upon which its constructions depend. Bodies can go to and come from the infinite but we compute their behavior in finite parts of space and only asymptotically with respect to the infinite.

It is different in projective geometry. There parallels do meet at infinity, and projective geometry underlies the perspective design and the corresponding theories of our view. Today we can formulate conceptually the difference between the two geometries; at that time this was not possible, there was only an idea, and it is natural that an artist could grasp it before and better than anyone else. This is why Raphael underlined with all means at his disposal this convergence towards the vanishing point in the open door: everything leads you to the infinite. Projective geometry and perspective serve here as a symbolic construction!

This is Raphael's first theological statement, and it is expressed through the mathematics of perspective. I doubt that any theologian could have expressed this idea; what was needed here was a mathematician, since mathematics is the very science of the infinite! But this statement was in line with the Florentine Platonism of the time, one of the roots of which was the *Docta Ignorantia* of the German philosopher and cardinal, Nicolaus Cusanus.

That I am not imagining these things is attested to by Raphael's second theological statement, and this time it is the architect who makes it, by designing this magnificent edifice. Namely, you look towards God, if you look towards him through the church.

Altarpieces of this time are loaded with theological implications, but are most often clad only in traditional, historic and sometimes accidental symbols. A construction like the one we find here, where architecture, mathematics and theology are so closely knit and intertwined, is surely extremely rare, if indeed not unique. For producing it, an architect, mathematician and theologian in one person was needed.

Acknowledgments

About 50 years ago, as a student 1 could regularly discuss the secrets of the use of perspective by the artists with my friend L. Burckhardt, and later, about forty years ago, I had the chance to have conversations on these questions with Erwin Panofsky. I discussed these questions with Dr. Mario Howald-Haller, Dr. and Mrs. Th. Beck helped me with the literature, and Mrs. M. Messmer pulled me out of laptop difficulties. 1 am greatly indebted to Kim Williams for translating my bad English into excellent American and for drawing the figures! Last but not least 1 am indebted to my wife for linguistic advice and for proofreading.

In Memory of Clifford Ambrose Truesdell and Edoardo Benvenuto

Introduction

The possibilities for learning the craft of the historian of science were hardly existent when I was a student; even today they are not abundant. In my case I learned what I could through a few, albeit very important, personal contacts, and one of the teachers I honour here.

But then I was also an avid visitor of art museums, and I read books and followed lectures on the history of the fine arts. One contact between the history of science and the arts themselves is immediate: often we can note that works of art prove that precise scientific and technological knowledge already existed at a very early time, long before a continuous and systematic development of science had begun. This knowledge was acquired by artists and artisans intuitively when they exercised their craft. Such testimonies are an essential part of the prehistory of a certain field of science, and the history of that field would not be complete without an account of them.

One example are the pyramids built during the old Egyptian empire. They testify to the powerful technology of their builders, but also to a geometric ideal shared by the sculptors of the same period. This must be seen together with how the statues express a new appreciation of three-dimensional space and of plasticity as an aesthetic value in itself. We may even see the spherical wigs that the ladies of that time wore as a compliment to that artistic ideal. Towards the end of the second millennium, we find the famous ornaments of the new Egyptian empire analysed by Andreas Speiser from the point of view of mathematics.[2] He pointed out that all tetragonal space groups had been found by the Egyptians, including one denoted C(4v)II, which is a highly nontrivial one, a fact that testifies to a systematic occupation with these geometric questions. This art was later developed further by the Arabs.

Another example are the ornaments of some circular medallions found in the so-called Tomb of Agamemnon in Mycenae, from the end of the sixteenth century B.C.[3]

---

[1] Originally published in *Essays on the History of Mechanics. In Memory of Clifford Ambrose Truesdell and Edoardo Benvenuto*, Antonio Becchi, Massimo Corradi, Federico Foce, Orietta Pedemonte, eds. (Basel: Birkhäuser, 2003), pp. 235–249.

[2] See Andreas Speiser, *Theorie der Gruppen von endlicher Ordnung*, 2nd ed. (Berlin: Julius Springer, 1927), pp. 87–96.

[3] See "The Symmetry of the Ornament on a Jewel of the Treasure of Mycenae" in this present volume, pp. 1–8.

This well-known figure, composed of six leaves, shows that its artisan had grasped, at least intuitively, a geometric theorem, perhaps the first one discovered in Europe.

Nothing needs to be said of the role the arts played in the genesis of Greek science and that of the artists of the Italian Renaissance in modern science: how many artists of those periods were not scientists and vice versa?

But there remains the puzzling question of the state of science in the Middle Ages, as testified by the great medieval buildings, especially the cathedrals and their towers, the highest towers built before the middle of the nineteenth century. That these works of art testify to an extraordinary engineering skill on the part of their builders is manifest, but is the beauty of these monuments also connected to their science? I mention this only because I shall return to this question later. But first I will address myself to the relation between historiography of the arts and historiography of science, which is my main theme.

Quite late in my activity as a historian and only slowly, very slowly, did it dawn on me that in spite of the fact that science and the fine arts are radically different endeavours of the human mind, the ways and means by which their historiographies proceed are often remarkably similar. And for reasons that are, for many, obvious, but which I hope will become clear to all in this paper, this symposium seems the appropriate place to consider these parallels.

I shall, of course, restrict myself entirely to the historiographies of the fine arts and the sciences; only at the end shall I say a few words about science itself.

The task of the historian of science

For his work, the historian of science first of all finds before him publications of the scientist, his articles and books, which present the scientific discoveries and also the deepening comprehension of earlier ones, both of which constitute the development of science. Behind these documents the historian can feel and sometimes even see, more or less clearly, the scientist who wrote them. Sometimes he also can see the context or line of tradition in which the scientist stands: his teachers, the books and articles that he read. More often he knows the disciples, students or other readers of a scientist's works, who formed what is sometimes called his "school," and more generally the historian can often discern the scientist's "sphere of influence," which expands for a certain time and then shrinks again. Sometimes it is more appropriate to focus on and speak of a team, today more so than ever. But teamwork has a long tradition in science, especially in the applied sciences: one has only to think of the architects and the builders of bridges, fortresses and ships, etc. These activities proceed side by side with interaction with developments in other, related fields, arithmetic, geometry, mechanics, chemistry, architecture, engineering, etc.; some of these interact with each other quite intensely. Sometimes research, especially in an applied activity, is initiated by the scientist but more often it is commissioned.

In short, we would say that the historian of science works with:

- the works of science;
- the scientists themselves;
- their teachers and disciples;
- the colleagues and contemporaries with whom they collaborate;
- interaction with other related disciplines;
- investigations initiated by the scientist or commissioned of him by a third party.

To each one of these points we find its counterpart in the historiography of the arts, namely:

- the works of art;
- the artists themselves;
- their teachers and masters as well as their disciples in the "atelier;"
- the team, especially in the architectural firm;
- progress in the various branches of the fine arts: architecture, sculpture, painting, decoration, and so on;
- commissioned works of art versus those that are due to the initiative of the artist.

But we must note that all these aspects and approaches were already being used by the historians of the fine arts, and very systematically so, at a time when there were only very few historians of science.

The significance of commissioned works

To begin with the last point, I think it is obvious that the great majority of works of art from the pyramids until at least 1800 have been commissioned: churches, altarpieces, palaces, villas and their decorations. Many of them must be counted as great works of art. Furthermore, monuments and frescoes were mostly not created on the initiative of an artist, but were, and still are, commissioned by a patron, and every art historian is aware of this.

Much the same holds for the development of science, but here the importance of commissioned research is in general not stressed enough. Again, think of the pyramids and the impact which their construction must have exerted on the then still rudimentary geometry and mechanics; think of the construction of the medieval cathedrals and towers; recall Galileo's praise of the Arsenal of Venice at the beginning of his *Discorsi*. But even in modern times this impact is often underrated; I remind you of cartography, the making of precision clocks, river corrections, the chemical industries and of the great laboratories during the second World War: Los Alamos, the Radar Lab and the Sonar Lab. Clearly all of these had a decisive impact in many ways not only on applied but also on fundamental science. I need not remind you of today's enormous efforts in the field of medical research, be they now commissioned by the state or guided by an enterprise such as a pharmaceutical company. These are all decisive pieces of the history of science and often very potent stimuli, but again, they are all too often underrated.

Relationships between domains of science

Another point where we can learn from our colleagues in the fine arts concerns the relationships between the various domains of science. To show this I will use examples of my own science, physics. It often strikes me how little aware many physicists and even historians of physics are of how much physics owes to its sister sciences, mathematics and chemistry.

That modern physics is inconceivable without the discovery and invention of the infinitesimal calculus has often been stated, but even so not always enough emphasis is given when one discusses the details of the progress made. If a scientist is classified as a mathematician, his contribution to physics may be undervalued, even if in absolute terms his contributions to mechanics and to physics fill far more volumes than his contributions to mathematics. I am speaking here of Euler, whose contributions to physics are slighted or even overlooked altogether by physicists and their historians. Much the same can be said of the work in mechanics of Jacob Bernoulli and of Cauchy. We needed Clifford Truesdell to show and explain to us the fundamental importance for mechanics and physics of the works of all three of them. On the other hand, this is a common problem. I am often surprised to see how much mathematicians and *their* historians underrate the stimulation their science received from other branches.

Likewise, historical accounts of physics often overlook the fact that, after Dalton, the real champions of atomism were the chemists: think of Kékulé and those who created stereometric organic chemistry, at a time when only a minority of physicists believed in atoms; some physicists even opposed atomism violently as late as the beginning of the last century. I need not say how little of all this arrived into the mainstream of the historiography of physics.

Now compare this state of affairs to a historiography in the fine arts, say a book about sculpture in the Renaissance. If this book did not present and discuss carefully the influence of related fields such as painting and architecture, it would be rated as a very poor book indeed.

Placing the scientist in his context

What is the state of affairs of our efforts to determine the intellectual roots of a scientist, the sources of his knowledge and, even more important, the efforts to find out what mainly stimulated him? And what is the state of our knowledge about a scientist's influence on those who came after him? In these cases I feel that we may justly claim that the historians of the arts have an easier time of it than we do. Why?

This, I believe, is mainly due to the overwhelming number of original scientific publications – books, papers, letters, and so forth –, which makes it extremely difficult to keep track of the transmission of knowledge through the generations and which makes *bona fide* errors almost unavoidable. Here is an example: I was told, or perhaps I myself concluded from what I was told, that Newton had learned Kepler's laws from the *Astronomia Nova* and from the *Harmonice Mundi*. Then almost accidentally I learned from I. Bernard Cohen, the well-known Newton scholar, that Newton had never read these books, but had learned the laws from a minor publication. Similarly, Newton in all probability never saw the *Discorsi*.

Thus, to show that someone had read a certain work of a predecessor is never as easy as it looks, and it is often impossible to prove that, even if he had read a work, he received an idea from it. Surely the way the art historians can document these links is impressive, but I would not be so bold as to tell a young historian of science that to find these links is a foremost priority for serious research. I would warn a young scientist to be careful with his time, interesting and even haunting or tantalizing as the question may be.

The historian and his object

But let me now come to the most important, indeed, the central comparison in this list of parallels between the historiographies of the fine arts and the sciences: the comparison between their respective relations to their immediate object of research, the work of art on the one side and the scientific publication – or, as the case may be, the experimental setup – on the other.

Historiography of the arts is inconceivable without an intimate, not only a visual, relationship to the work of art, and I suspect that in most cases such a relationship must mature over a long period of time. Every art historian knows that the works of art and their contemplation are the *raison d'être* of his whole profession; indeed, for most of them it is the principal motivation for their intense and often passionate relationship with the history of the arts. With respect to the historiography of science, we can observe over and over again that this is not necessarily the case at all.

Of course, the art historian has it much easier here ... at first: works of art are directly accessible to him, just as they are to everybody, especially in museums but even on our public streets and squares. Therefore it is also easier to penetrate into them and to gather facts concerning them. It is precisely this "easy access" that the historian of science lacks. The older scientific documents are not easy to find, let alone to understand, as everyone who has tried will agree. Thus for understanding and inserting into their proper historic context the old as well as the new documents, one must have been educated and trained in the specific scientific domain being studied, whereas the art historian need not be a professional artist.

For this reason alone I am convinced that the most important task in the field of the history of science is to edit the old documents and to make them accessible for scientists through introductions and comments. I want to underline here that both Clifford Truesdell and Edoardo Benvenuto did this in an exemplary way.

Here an important lesson awaits all scientists who care for the history of their own field, and thus I shall appeal to them. It is inconceivable that anyone should write about an artist without having thoroughly looked at his works and having pondered them. The art critic Souren Melikian wrote, on the occasion of an exhibit entitled "Rembrandt's Women," that while, except for the three or four women most closely connected to the painter, little was known about the women whom he painted, a careful look at the faces in the portraits and etchings would reveal Rembrandt's great psychological sensitivity for each individual and yield to the spectator deep insights of each one.[4] And indeed the same holds true for scientists.

The physicist Arthur Wightman once told me that André Weil, who, while writing his book *The History of Number Theory* studied the works of Euler extensively, said in a private lecture, "after having now penetrated into Euler's work on number theory, I think that I know him better than I know most of my best friends!"

Even the writer of a *biographie romancée* that is only flimsily connected to its subject, takes a good, if perhaps mostly sentimental look, at the works of his artist. But all too often, I am afraid, this does not hold for the scientists.

I am sure that all of you have heard anniversary lectures delivered by a scientist, perhaps even at a university, where the speaker had manifestly not bothered to lift the cover of even one book of the famous man whom he was invited to honour. Rather he was satisfied to tell his audience what he had learned from the footnotes he had found in the textbooks read during his student years. Or perhaps he had heard it only in one of the lectures he attended (his own?). To act so with respect to scientific matters proper is inconceivable, but concerning the history of science everything seems to be permitted – and accepted!

Here I must appeal to all historians of science to do all in their power to prevent such unworthy performances. They are not only indecent, but they do great harm to the cause of the history of science as a rigorous academic discipline, to both its research and its teaching: in short, they are counterproductive! Small wonder that so often the history of science is considered a mere curiosity. You know that there are many universities where history of science is not a regular discipline, and if such fraudulent and swindling speeches are given, the authorities, and even more

---

[4] Souren Melikian, "The Unsolved Riddle of Rembrandt's Women," *The New York Times,* November 3, 2001.

importantly, our dear colleagues, can say, "why subsidize chairs and even institutes, when each professor can do this for free?"

That is not all, of course. The history of science claims, rightfully I think, to broaden the horizons of its students and of all interested persons, scientist or not. And again it is not the least merit of Truesdell and Benvenuot that they did just that. Thanks to both of them we have now a history of science richer than ever before, and which does indeed open new horizons.

Before saying a few words about the men we honour today, I must add a point, a main point indeed, thanks to which the historian of science often has it easier than his colleague in the arts. With respect to the arts, the German historian Ranke's dictum that "every epoch is immediate to God" is valid;[5] there is no progress. While I am not saying that after the Italian Renaissance no equally great works of art were produced, I am certain that during later periods, no works *greater* than the most beautiful ones of the Cinquecento were ever made.

Not so in science. There we can notice and establish progress. By this I do not mean the mere accumulation of vaster and vaster amounts of detailed information, let alone a general increase of the quality of research. Rather, through time science establishes more and more connections between more and more phenomena in each field and, through the discovery of new concepts and new laws, eventually creates larger and larger, connected domains. Especially in the exact sciences, thanks to newly-discovered and rigorously-developed concepts and to theorems formulated in mathematical language, what we have learned by observation, or experiment, by imagination and deduction becomes unified. These unifications of the exact sciences through general theories are, I believe, the best measure of this progress. This stands in striking contrast to what we see in the history of the arts, with its upsurges and downfalls, where progress exists only in a subordinate and relative way.

As a consequence of this progress through unification the historian of science can always say to even the greatest scientist who preceded him, "while I admit with great respect and pleasure that you penetrated more deeply into nature than I do, I can now see more clearly than you could, what you were up to!" No art historian, nor even an artist, can say this. This possibility, which is based on the new perspectives offered by the progressing unification, makes our task easier. It is also a great relief for the historian; it is certain that in the not-too-distant future, historians will see many things even more clearly than we can see them today.

It follows that history of science will never give a final and absolute account; its account can always be improved upon. Thus the historian must be judged according to the words of the German poet Friedrich Schiller, "He who has done his best for his own time has lived for all times."[6] It is in this spirit that we must honour Clifford Truesdell and Edoardo Benvenuto, about whom I shall now say a few words

Benvenuto and mechanics

I will begin with Edoardo Benvenuto, whom I knew much less well: once I was his guest and he showed the beautiful churches and *palazzi* of Genoa to my wife and me! But except for the conversation during the meal to which he invited us, most of

---

[5] Editor's note. The full quote is, "I would maintain that every epoch is immediate to God, and that its value consists, not in what follows it, but in its own existence, its own proper self." Leopold von Ranke, *The Secret of World History: Selected Writings on the Art and Science of History*, edited and translated by Roger Wines (New York: Fordham University Press, 1981), p. 159.
[6] Friedrich Schiller, *Wallenstein's Camp* (1798), Samuel Taylor Coleridge, trans.

what I know about his views and ideas I gathered from copies of papers sent to me by Profs. Foce and Corradi, to whom I am much indebted. I learned from at least two of these works that our ways were strangely intertwined; we dealt at least twice with the same subject and both times each of us was ignorant of the other's efforts, which now, of course, I regret.

The first subject with which we both dealt, I on my side in collaboration with Patricia Radelet-de Grave and J. L. Pietenpol,[7] was Daniel Bernoulli's first paper on mechanics, where he deals with the "parallelogram of forces." This paper opened a long series of axiomatic, as we say today, investigations on the law of the composition of forces, thus on the concept of force. Benvenuto carefully follows the long story of these investigations, from which much can be learned.[8] Mrs. Radelet's and my own aim, when, some time later we edited the third volume of the complete works of Bernoulli, was more modest: we just wanted to explain in detail what he did and why, and indicate some later criticisms. I regret, not so much that I could not discuss Benvenuto's and our *answers*, but that I could not discuss the *questions* to be discussed here. For as Clifford Truesdell never tired, and rightfully so, of stating, Newton's concept of force, which he called his greatest creation, is the first cardinal point of mechanics. The parallelogram law is the mathematical side of this concept, so that this question has a deep philosophical significance too. I would love to have heard Benvenuto's opinions about this.

But then Prof. Corradi sent me another article by Benvenuto, one even more intriguing and fascinating, entitled "L'ingresso della storia nelle discipline strutturali."[9] This paper comes very close indeed to accomplishing the task of finding in the history of the arts testimonies to the prehistory of a certain field of science.

In the background of Benvenuto's discussion stands the relationship between science, as represented by mechanics and engineering, and the arts, as represented by architecture. To be sure, he has here, above all, practical problems in mind, problems of restoration of old buildings. In his own words,

> The greatest push came from ... the need to formulate plausible interpretations of the mechanical behaviors of structures and construction materials which had accompanied human civilization for millennia, such as wall structures, vaults, domes.[10]

So, like me, he was trying to find a bridge between the arts and science, but then he was looking at this bridge from the other end as well, as for example when he speaks "of the choice made by [the journal] *Palladio* to integrate History of Architecture with the History of the Sciences which has always supported the art of building."[11]

---

[7] David Speiser, "Bernoulli, Mechanics and Restoration," *Die Werke von Daniel Bernoulli,* (Basel: Birkhäuser, 1987), vol. 3, pp. 6–23.

[8] Edoardo Benvenuto, "The parallelogram of forces," *Meccanica,* vol. 20, no. 2 (1985).

[9] Edoardo Benvenuto, "L'ingresso della storia nelle discipline strutturali," *Palladio* (Nuova serie) no. 1 (1988): 7–14 .

[10] *La maggior spinta è venuta ... dall'esigenza di formulare plausibili interpretazioni del comportamento meccanico di strutture e materiali da costruzione che da millenni accompagnano la civiltà umana, come le murature, le volte, le cupole* (Benvenuto 1988, *op. cit.,* p. 7).

[11] *...della scelta operata da* Palladio *per una integrazione della Storia dell' Architettura con la Storia delle Scienze che da sempre hanno sostenuto l'arte del fabbricare* (Benvenuto 1988, *op. cit.,* p. 7).

The cardinal point of the paper is a long quotation from the work of the German philosopher Arthur Schopenhauer.[12] He makes a series of comments in support of the philosopher's thesis:

> The struggle between weight and stiffness constitutes ... the only aesthetic theme of art in architecture; to throw light on this struggle in the most various and evident ways: this is its office.[13]

And further,

> ... braking [the forces] by deviating them; this prolongs the struggle, and makes visible in a thousand various forms the untiring efforts of the two enemy forces. If left to its natural tendency, the entire building would form a compact mass that pressed ... on the ground, to which the weight pushes it inexorably. ... Instead, the stiffness ... opposes this force with an energetic resistence. ... Thus the beauty of an edifice consists in the evident final adaptation of each part.[14]

Benvenuto then analyses what he calls

> ...the "aesthetic" traces of the three principal themes of structural mechanics: the first is ... the "struggle" between the "natural tendency" and the energetic resistance; the second ... are ... the "tortuous pathways" by which architecture offices a mediated manifestation of gravity; finally, the third is .. the "immanent finality" that directly "refers to the statics of the whole", conferring coherence and teleonomic harmony to every single part.[15]

From this basis Benvenuto sketches a program, especially for architectural archaeology, but as I am not competent in this field, I shall not follow him further in this direction. However, what strikes me, although Benvenuto does not ask it himself, is the following question: Which period and which style are evoked by Schopenhauer's conception and Benvenuto's analysis? Surely neither the Greek nor the style of the Renaissance, where combinations of symmetries, distinguished geometrical figures, proportions and so forth, led to the harmonious, ideal building. Nor is it the Baroque, where such a *lotta*, or struggle, is usually carefully concealed,

---

[12] Benvenuto quotes from Arthur Schopenhauer's *Die Welt als Wille und Vorstellung* (Leipzig 1819), vol. III, p. 43.

[13] *La lotta fra il peso e la rigidità costituisce ... l'unico tema estetico dell'arte in architettura; far risaltare tale contrasto nel modo più vario e più evidente: questo è il suo ufficio* (Benvenuto 1988, *op. cit.*, p. 9).

[14] *... frenando [le forze] col deviarle; così prolunga la lotta, e rende visibile sotto mille forme svariate lo sforzo infaticabile delle due forze nemiche. Abbandonata alla sua tendenza naturale, tutto l'edificio verrebbe a formare una massa compatta premente ... sul suolo, su cui lo spinge inesorabile il peso ... La rigidità invece...oppone a tale sforzo un'energica resistenza... Quindi la bellezza di un edificio consisterà nell'evidente adattazione finale di ogni parte* (Benvenuto 1988, *op. cit.*, p. 9).

[15] *...la traccia "estetica" dei tre principali temi della mecanica strutturale: il primo è ... la "lotta" tra la "tendenza naturale" e l'energica resistenza; il secondo ... sono ... le "vie tortuose" per quali l'architettura offre una manifestazione mediata della gravità; il terzo infine è ... la "finalità immanente" che direttamente "si referisce alla statica dell'linsieme" conferendo coerenza e teleonomica armonia ad ogni singola parte* (Benvenuto 1988, *op. cit.*, p. 9).

for instance, by refined artifical painted architectures. It was with these two styles that, naturally, Benvenuto was mostly concerned.

But Schopenhauer's words evoke the Gothic architecture of the Middle Ages, especially in France and Germany; indeed, the Gothic is the only style explicitly named by Schopenhauer in his text. I often wondered which geometric laws play a decisive role in Gothic architecture, the way they do in other styles. No doubt there are such mathematical laws in the Gothic style: just look at a cathedral and especially its plan, which is the result of an enormous imagination, punctiliously organized, even in the smallest details. And what is mathematics but organized imagination? Benvenuto's article suggests this to me, although he does not say it: it is not geometry, but her sister (or, if you prefer, her daughter) mechanics, that plays a decisive role here. I prefer not to say "statics" and would even prefer to say "dynamics," in the sense of the science of forces (*dynamis*) in equilibrium. And lest I should be misunderstood, here I do not mean "dynamic" in a vague philosophical sense. Exactly as "geometry," when we speak of the art of the Renaissance, means the geometry of the textbooks, in this instance I mean the mechanics of the textbooks: the science not only of the shapes of bodies, but of their interaction. We know from Truesdell that, during this period, problems of resistance and elasticity were investigated. I shall quote him in a moment.

But in place of circles, squares, and proportions of figures, of what exactly must we think here? How do the dynamic ideas guide the builder and materialize into a beautiful building?

It is here that Truesdell comes in. He showed that Jordanus de Nemore had conjectured – if erroneously so, as we know from Jacob Bernoulli – that the *elastica*, that is, the curve described by a bent beam, is a circle.[16] Now, several people have told me that the Gothic pointed arch is composed of two circles. But aesthetically these circles do not serve the same purpose as in classical architecture: they are not meant to be perfect *geometric* curves, but rather perfect *mechanical* curves. And thus Truesdell assures me, that what Benvenuto suggests to me goes in the right direction!

These are the curves that guarantee, as the pessimistic Schopenhauer might have put it, "for a building, whose height is dictated, the greatest possible security." A mathematically-equivalent, yet optimistic formulation, which expresses, even better, the aim and boundless ambition of the builders is "given a dictated amount of money, material and thereby guaranteed security, these curves permit building under these conditions the highest possible cathedral." Hence the strong vertically-thrusting drive that we observe and experience in a Gothic cathedral.

It would be lovely to continue this dialogue with Benvenuto on the *estetico*, but that would necessitate more space than is allotted to me, and frankly, as you guess, at the moment I am not yet fully prepared for this.

Before going over to Clifford Truesdell, let me add one more quotation from Benvenuto's article, about the role of historiography:

> Perhaps it is not so much the historiographic objective that holds this interest, as much as the awareness that today a more profound knowledge and a careful remediation of the past are necessary conditions for a real progress in research.[17]

---

[16] See Clifford A. Truesdell, *The Rational Mechanics of Flexible or Elastic Bodies, 1638–1788, Leonhardi Euleri Opera Omnia*, Series II, vol. 11b (Zurich: Orell Füssli, 1960), p. 420.

[17] *Forse non è tanto un obiettivo storiografico ciò che sostiene questo interesse, quanto piuttosto la consapevolezza che un'approfondita conoscenza e un'attenta rimeditazione sul*

Truesdell and historiography

When Clifford Truesdell entered the field of history of science, he was already the well-established author not only of many publications in various fields, but especially of many handbook articles. Furthermore, together with Walter Noll, he had formulated what may be called a new, deeply-structured mechanics of continua, which in their hands, through the use of Hilbert's axiomatic techniques, had become a totally unified domain.

This proved to be the best conceivable preparation: equipped with this powerful armour he now turned more and more to the history of classical physics. Indeed, he could now see and display the historic development in perspective, step by step, the new ideas, the fruitful ones and the failures, the stimuli that they exerted, the accomplishments, always presenting clearly not only the context in which each scientist worked, but also "what he was up to." Combined with an intimate acquaintance with the original sources, he was able to assign to every question the exact role it had played, as well as to every author his proper place in that big stream. He showed how it all came together into what we know and understand today. Thereby, he was often able, with just a few strokes, to sketch the portrait of each actor in this epic and make visible the merits of the great ones as well as the forgotten, overlooked, and unlucky ones. Above all, he could display the greatness of the field, whose history he was telling. I especially recall here his Introductions to Euler's work in hydrodynamics and in the theory of elasticity.[18]

The introduction to Euler's hydrodynamics arrives at its summit when Truesdell shows how, after Newton's work, Daniel Bernoulli through his equation unifies hydrostatics and hydraulics; Johann I Bernoulli applies Newton's concept of force to fluids; d'Alembert reformulates this science by a field description and, together with Euler, introduces partial differential equations into this domain. Eventually Euler, after having recognized the importance of what we call today an "inertial system," creates the new central notion: the "inner pressure." These discoveries allowed him to formulate the "Euler equations," the first field theory to unify hydrodynamics and aerodynamics! And thereby, as Truesdell showed, he opened the way to Cauchy's definite formulation of the theory of elasticity.

The introduction to Euler's works in the field of elasticity is a monumental treatise that, in its main part, traces its subject from Galileo to Coulomb. In the Prologue, which sketches its prehistory of more than 2000 years, he writes about the Middle Ages, and especially about the book written during the Gothic period entitled *Theory of Weight* by Jordanus de Nemore, who I mentioned earlier:

> ... remarkable it is, Western in spirit, and ambitious beyond anything in the Greek and Arab tradition. The seventeen propositions on fluid flow, resistance, fracture and elasticity are all original....[Jordanus's] attempt at a precise argument to prove a concrete result in a domain never previously entered is of splendid daring.[19]

---

passato sono oggi condizione necessaria per un reale avanzamento della ricerca (Benvenuto 1988, *op. cit.*, p. 11).

[18] Clifford A. Truesdell, "Editor's Introduction: Rational Fluid Mechanics, 1687–1765," *Leonhardi Euleri Opera Omnia*, Series II, vol. 12, pp. VII–CXXV (Zurich: Orell Füssli, 1954); and *The Rational Mechanics of Flexible or Elastic Bodies, 1638–1788, Leonhardi Euleri Opera Omnia*, Series II, vol. 11, part 2 (Zurich: Orell Füssli, 1960).

[19] Truesdell 1960, *Rational Mechanics, op. cit.*, p. 18.

The main story is long, complicated and intertwined, and I cannot go into it here. It may suffice to say that, thanks largely to this Introduction, the history of the theory of elasticity, although the most complex part of the mechanics of continua, has become today perhaps the most carefully investigated one.

You will have noticed that often when I speak about Truesdell I use the word "concept." Indeed, he never tired of stressing the importance of the concepts used in science, especially of the concept of force, and it is from him that I learned, alas only very late in my career as a teacher, the importance of the role played by concepts. The concept is the mediator between the world of mathematics and the world of the senses, which in the end makes a science of the laws of nature possible.

This role is displayed and discussed in the book that Truesdell wrote with S. Bharata, *Classical Thermodynamics as a Theory of Heat Engines.*[20] This is certainly his most important contribution to the art of teaching, from which every teacher can profit. It should be mentioned that Truesdell stresses here that for each theory presented to the students the teacher must indicate from the start the limits that the basic assumptions impose on it.

Conclusion

The reader may now ask, "What did the two men, Clifford Truesdell and Edoardo Benvenuto, have in common?"

First, both were great teachers, if we consider this word in its largest sense. I will only mention here some of the things that resulted from their parallel activities. Truesdell's work as a historian, especially of the seventeenth and the eighteenth centuries, is continued in two recent books written by Giulio Maltese on the history of mechanics from Newton to Euler, written in part in Genoa in close contact with Benvenuto.[21] These two books, whose author has become a foremost scholar in his field, are, I suspect, the first ones that profit fully from Truesdell's work and are a continuation of it.

Truesdell's health did not permit him to edit parts of Daniel Bernoulli's works, as he had hoped, and this task passed to Prof. Gleb Mikhailov. But the Bernoulli Edition, as it stands today, is unthinkable without his ideas, his experience, his unfailing advice, also as regards the beauty of the volumes and, most importantly, his constant encouragement. Thanks especially to Patricia Radelet-de Grave, the school of Genoa founded by Benvenuto is now, together with Prof. Maltese, collaborating with this enterprise, so that the efforts of both men do come together here.

And finally, both men were great lovers and connoisseurs – indeed, experts – of the arts. No need to remind anyone in Genoa about this of Edoardo Benvenuto. But I may add a few words about Truesdell, for in this field too his knowledge and erudition were stupendous. In Baltimore he had many friends among a group of craftsmen, who worked with him on the decorations of his stately home, where both he and his wife Charlotte, herself a musician, organized concerts played by artists on instruments built after the ones used in his beloved eighteenth century, the "Age of Reason," as he lovingly called it.

---

[20] Clifford A. Truesdell and Subramanyam Bharatha, *The concepts and Logic of Classical Thermodynamics as a Theory of Heat Engines: Rigorously Constructed upon the Foundation Laid by S. Carnot and F. Reech* (New York: Springer-Verlag, 1977); see also pp. 110–111 in this present volume.

[21] Giulio Maltese, *La storia di "F = ma". La seconda legge del moto nel XVIII secolo* (Florence: Olschki, 1992); and *Da "F = ma" alle leggi cardinali del moto: sviluppo della tradizione newtoniana nella meccanica del '700* (Milan: Hoepli, 2002).

Truesdell had a great sensibility not only for music, but also for the beauty of scientific texts, admiring the organisation of the matter presented, the transparency of a proof, and the beauty of the mathematics or the mechanics itself, especially in the works of Euler, who he compared to Bach and Mozart.

No wonder then that, as Charlotte Truesdell told me, Benvenuto and Truesdell knew, liked and respected each other, and that Truesdell (and knowing Mrs. Truesdell as I do, she as well) assisted Benvenuto with the English translation of his book.

Thus both men whom we honour here taught us that science, if presented by great men, can be beautiful too. When presenting its history this is much harder, but again both Benvenuto and Truesdell show us that this is possible: may their example excite and guide many followers!

Acknowledgments

It is a pleasure to thank Prof. Massimo Corradi as well as Prof. Antonio Becchi and Prof. Federico Foce for their kind invitation to participate at the International Symposium "Between Architecture and Mathematics: The Work of Clifford Ambrose Truesdell and Edoardo Benvenuto," which took place in Genoa, Italy, 30 November–1 December 2003, and for their constant and generous hospitality. I am indebted to my wife for linguistic advice, and to Kim Williams who assisted in the edition of the *Proceedings*.

The Importance of Concepts for Science[1]

To Professor Piero Villaggio on the occasion of his 70th birthday

When I lectured at the Scuola Normale Superiore di Pisa, I regularly observed in the lunchroom a lean professor with a somewhat haggard face who always ate with a great appetite, but whose name I could not make out. But then the Dipartimento di Ingegneria Strutturale of the University of Pisa organized a lecture by Clifford Truesdell. Luckily I sat next to Charlotte Truesdell, who answered my question, "This is Villaggio," emphasizing, "He is very good!" She then introduced me to him, and since then we had more and more contacts. He even honored me by participating in my seminars. He became an editor of a volume of the works of Bernoulli, he helped me with an article on Galileo,[2] which he translated, and, best of all, a good friendship grew between us. For all this I am deeply grateful to Piero and in these lines will try to show it!

Introduction

Clifford Truesdell, one of Piero Villaggio's mentors and also a teacher of the writer of these lines, stressed again and again the importance of the concepts for mechanics – indeed, for science –, and he called the concept of force Newton's greatest creation. It is a fact that the importance of concepts is in general much underrated among scientists, and often much too little attention is paid to it by the teacher who introduces a new concept in a lecture. Thus, I propose to show here why concepts are not only important but indeed fundamental for the organization of a science. This holds not only for so-called pedagogical reasons, but even more for purely scientific and sometimes even philosophical ones.

As is well known, concepts were first systematically investigated by Plato, who called them "ideas" in several of his dialogues, notably in the *Parmenides*[3] and the *Sophists*. Today we often distinguish between idea and concept, keeping the former word for philosophy. But, what Plato says also covers what we today call, especially in the sciences, a concept. The use of the concepts is probably the point where science comes closest to philosophy. First I shall discuss a few examples which show that the use of concepts is inevitable, and I shall try to show the reason for this. Then I will try to pin down and explain, as well as I can, what the role and the function of a concept are in physics.

---

[1] Originally published in *Meccanica* (*International Journal of Theoretical and Applied Mechanics*), vol. 38, no. 5 (2003): 483–492.
[2] "The Beginning of the Theory of Elasticity from Galileo to Jacob Bernoulli," in *Galileo Scientist, his years at Padua and Venice*, Milla Baldo Ceolin, ed. (Padua, 1992), pp. 87–112.
[3] See Andreas Speiser, *Ein Parmenideskommentar. Studien zur platonische Dialektik* (Stuttgart: K.F. Koehler Verlag, 1937 and 1959).

PHILOSOPHIÆ

NATURALIS

PRINCIPIA

MATHEMATICA.

Autore *JS. NEWTON*, *Trin. Coll. Cantab. Soc.* Matheseos
Professore *Lucasiano*, & Societatis Regalis Sodali.

IMPRIMATUR·
S. PEPYS, *Reg. Soc.* PRÆSES.
Julii 5. 1686.

LONDINI,
Jussu *Societatis Regiæ* ac Typis *Josephi Streater*. Proftant Vena-
les apud *Sam. Smith* ad infignia Principis *Walliæ* in Cœmiterio
D. *Pauli*, aliosq; nonnullos Bibliopolas. *Anno* MDCLXXXVII.

Title page of Newton's *Principia*

My first example will be precisely Newton's force; I shall recall in a few words
Newton's procedure in the *Principia*, but then, taking to heart a piece of advice from
Truesdell, I shall follow a line initiated by Daniel Bernoulli, which avoids a difficulty
that is too often ignored in introductory lectures.

Why are concepts important in science?

For solving a problem in science and, even more so, for elaborating a theory and confirming it, many discussions by word or by letters between various people pursuing very different activities are needed. Some of them observe, some compute, others make experiments and still others try to elaborate a theory, as simple as this theory may be. But in order that the meaning of what is said in this discussion be understood exactly by the other participants, mere words and even mere formulae do not suffice. The appeal to "rigorous definitions" does not help much either, since definitions too are always only made up of other words.

Long before the modern development of mechanics began, the foundations of geometry and astronomy had already been laid. Perhaps the oldest scientific observations are those made on the positions of the stars. But how could an Egyptian astronomer communicate his observations to a colleague in distant Babylon, so that he may reproduce his measurements? Both must agree on what an angle is, and on how one measures it, and both must agree on the exact time difference between, say, sunrise at the Euphrates and at the Nile. "Angle" and "duration" are both geometric concepts, for which, if they are used by astronomers, a similar discussion ought to be made. This use implies that Euclidean geometry can be applied to nature, but here I will leave this alone.

However, while geometry deals only with the extension of figures in space, mechanics deals with the interactions of bodies in space and time. This endeavour also supposes that one has verified the validity of Euclidean geometry for the description of the figures of the bodies and their relative locations in the space in which we live, observe and make experiments.

Thus, for penetrating into a new dimension of our scientific experience, as mechanics indeed does, new concepts that lead beyond those of geometry and kinematics are needed. Kepler, Galileo, Descartes, Huygens and Newton realized this, and for classical physics the central and most powerful of them is Newton's force. Thus this example seems best suited for showing the precise role played by concepts in the so-called exact, or better, mathematized, sciences.

What is a Concept?

Like the Roman god Janus, a concept must look in two opposite directions. Since we insert a concept as a mathematical quantity into an equation, it must look towards mathematics, for we must know how to calculate with it, as Newton made it clear indeed: right after his famous three *Axiomata sive Leges Motus*, Newton proves as Corollary 1 the parallelogram law for the addition of two forces! Today we say briefly that the force is a vector; when we say "the force," we mean any force whatsoever. So we know how to insert any force into an equation, connect it with other concepts, and compute with it.

But in mechanics we use many other vectors: velocity, momentum, etc. Therefore we must indicate what distinguishes the force from all others. Obviously geometry and algebra cannot teach us this. Here we must look with the eyes of the other face of Janus, the ones that look to what we observe with our senses directly or through an instrument, and this is usually the experimentalist's starting point. What is needed here? Simply that we fix an unequivocal prescription, which says how a force must be measured. Usually this is done by a scale: we select a standard force, namely gravity, and then fix a standard unit, that is, some fixed weight. And with a clever supplementary device added to the scale, we can also measure other forces, which are electric, magnetic, etc.

Thus we can now truly speak of the force, whatever the origin of any particular force may be, for by such prescriptions for measuring the mathematical quantity the force is now also connected to our experience through the senses.

Thanks to the two-sidedness the concept of force is capable of connecting all interactions of one body with another, for it connects the world of mathematics with the world of our experience, gained through our senses. It is a bridge over the large abyss that separates these two radically different worlds. And, in doing so, last but not least, it also connects theoreticians and experimentalists, who work on the same problems but from different sides and yet must communicate. Thus concepts are irreplaceable for laying the foundations of mechanics, indeed of all sciences of nature.

"The proof of the pudding is in the eating," and this connection now permits scientists to calculate exact predictions, which an experiment can confirm, ... or perhaps contradict! Consider a system of two weights of 3 and 4 kg, respectively, linked by a rope that lies over two distant wheels. They are not in equilibrium, but if we attach a third weight of 5 kg between the two wheels, this system will be in equilibrium, and we can predict that, at the point where the third weight is attached, the angle between the two parts of the rope is exactly a right angle!

Of course, this power to predict can eventually also become, after a long chain of theories and experiments, the motor for the high altitude flights of the construction engineer's imagination!

What Newton said about the concept of force

When I sketched this little experiment, the critical reader will have noticed that while at first I had indeed followed Newton, I then left him, and that so far only the foundations of statics have been laid, of which Newton never speaks in the *Principia*. Rather, taking to heart the advice of Truesdell, I followed a line initiated by Daniel Bernoulli. So why not follow Newton more closely?

Newton's "Second axiom or law of motion," often believed to introduce the concept of force, is usually quoted today as "force equals mass times acceleration," and sometimes even as "force is mass times acceleration." What is wrong with these quotations? The answer is, almost everything!

First, Newton did not write mass times acceleration, but, in modern parlance, he wrote force equals the time-derivative of the momentum; he did this, as we shall see, for good reasons. Both these concepts are the subject of the first two of the eight definitions stated right at the beginning of the *Principia*. The definition of the mass was often criticized as circular, as Newton defines the quantity of matter as the product of its density times the volume, a procedure due probably to his atomistic views. But, however that may be, for satisfying the two postulates stated above, mass and momentum are best introduced by using the laws of collision, discovered by Wren, Wallis and Huygens, and also quoted in the *Principia*. These laws state the conservation of momentum in collisions, according to which for each body there exists a quantity – the mass – such that the sum of the products of each body's mass times its velocity is the same after the collision as before. An analysis of the law then also yields a clear prescription for measuring the masses and the momenta.

Second, one cannot introduce two new notions here, the mass and the force, with only one equation. While mass and momentum belong to the same body, the force does not belong to the body at all, and it will become clear that the second axiom refers to all forces whatsoever. Whence the immense power of this concept! I suspect that it is for avoiding this difficulty that Truesdell recommended introducing the force in a first step in the frame of statics only, where the mass is not needed.

The third error is the most crucial one: My guess is that nothing has contributed as much to the confusion of so many students of physics. Reading the beginning of the *Principia* carefully, one sees that the second axiom in fact says something very different from the "quotations" stated above.

Of the eight definitions mentioned before, the third and the fourth ones refer to forces. Definition IV says:

> The impressed force [*vis impressa*] is the action exerted on a body for changing its state of rest or of uniform, rectilinear motion.

Thus it covers what we call a force still today.

But what does Definition III tell us? It says:

> The force that is rooted in the (bodies) mass (*vis insita*), is the power to resist, through which each body, as much as it depends upon itself, perseveres in its state of rest or of uniform, rectilinear motion.

In his comment on Definition IV Newton also calls the *vis insita* the *vis inertiae*, adding that a body exerts this force only when its state is changed by another force impressed upon it.

So Newton is far from defining, let alone explaining, the force as "mass times acceleration," or even as the time derivative of the body's momentum, although these meanings were often taken as the content of the second axiom. Newton was too sharp a logician as well as too good a mathematician to ignore the fact that if the left-hand side in an equation is a force, then the right-hand side must be as well, for otherwise we have what the Belgians call *une addition de vaches et de cochons*.

Newton simply states as an axiom: for obtaining the equation of motion for a body that is propagated by a force, one must put the *vis impressa* equal to the *vis insita*; and that this *vis insita* is equal and opposite to the time-derivative of the body's momentum, defined in Definition II.

Here a remark is called for: this *vis insita* or *vis inertiae* is by no means what we call today an "inertial force"! The *vis insita* is a vector, a quality which our inertial forces lack; for instance the fact that "inertial forces" appear in an equation can be simply the consequence of not using Cartesian coordinates. Of this more below.

Newton's *vis insita* practically disappeared from writings on mechanics. Here I can mention only three scientists who paid attention to this problem.

Jacob Bernoulli based his theory of the pendulum, as Truesdell noted, on the balance of angular momentum. For his computation of the frequency of an oscillating, rigid body, he started from the second necessary condition for the equilibrium of a rigid body, namely, that at all points the sum of all moments of forces that act on this point, must be zero. But if the body oscillates, one must add for each point to this sum, what Newton, whose *Principia* Jacob did not know, might have called "the moment of the *vis insita*."

Jacob's procedure is complicated, since he did not yet know two key concepts: the angular velocity, a skew tensor that makes the use of polar coordinates transparent, and Daniel Bernoulli's moment of inertia, which makes it possible to integrate over all points right from the start. Since Euler, who generalized the motion of inertia to the three-dimensional, symmetric tensor of inertia θ, this law of motion can simply be written as: $\theta\dot{\varpi} = M$.

In 1752 Euler introduced "a new principle in Mechanics."[4] It was Newton's second law, but written in Cartesian coordinates, and in what we call today an "inertial system." That he knew what he was doing is clear from the discussions of the principle of relativity in several of his works. A fruit of this new principle, as we shall see, was his equations of hydrodynamics.

This development is little known, but fortunately today we have, besides Truesdell's, the two books by Giulio Maltese, where the historical development from Newton to Euler is carefully presented, and where excellent use of Truesdell's historical work is made.[5]

And then we have Einstein's so-called theory of general relativity, which ought to be called rather "theory of the inertio-gravitational field." Einstein made of Newton's rigid space-time an elastic one: the metric field.

But in spite of this enormous step ahead, one cannot say that we have understood the *vis insita*: it remains the greatest and probably deepest puzzle of classical, as well as quantum physics, where it is also tied to the so-called mass-spectrum.

The other concepts

What was done here for the force must be done for each concept used in physics: first the rules for calculating with this quantity must be laid down; this is done usually with the help of tensor algebra or as in the mechanics of continua, by tensor analysis. Why just these two? This is a consequence of the principle of relativity, which says that a group – the Galilei or, according to Poincaré and Einstein, the Lorentz-Poincaré group – acts on space and time. In particular the elements of the group transform one "inertial frame" into another (of these we shall hear more), and under such transformations the equations remain invariant.

But then for each concept a prescription for measuring this quantity must also be given, and for satisfying this second postulate one needs a new, often highly sophisticated instrument. Thus, for extended, rigid bodies, besides the force, one needs the concept of "moment of a force." This quantity, unlike the force, is a skew-tensor, and for measuring it one needs a Roman scale rather than an ordinary one.

If we turn to deformable bodies, that is, to fluid and elastic bodies, we must extend tensor algebra to tensor analysis. The first new concept that we need is a purely mathematical one, namely, that of a "field." This is essentially a tensor defined at all points of space, or possibly of space-time.

The simplest fields are the density-field of a body and the so-called inner or scalar pressure introduced by Euler. This concept is often confused with the outer pressure introduced by Simon Stevin. The difference between these two concepts is a good illustration of what was said before, as it touches on both postulates which they must satisfy. Namely, the inner pressure is simply a scalar, or more precisely, a scalar density, while the less often used outer pressure is, with respect to tensor calculus, something much more complicated. It is the ratio of the force by a surface, that is, of a vector by a skew-tensor, and this makes it unwieldy when used in a computation.

But the two concepts differ even more with respect to the second postulate: how are they to be measured? Here, at first, the outer pressure seems to be superior: force and surface are both completely measurable quantities and so is therefore their

---

[4] Leonhard Euler, "Decouverte d'un Nouveau Principe de la Mecanique" (1752), E 177. *Mémoires de l'académie des sciences de Berlin* 6, 1752, pp. 185–217; *Leonhardi Euleri Opera Omnia*, series II, vol. 5, pp. 81–108.
[5] Giulio Maltese, *La storia di 'F = ma': La seconda legge del moto nel XVIII secolo* (Florence: Olschki, 1992); *Da 'F = ma' alle leggi cardinali del moto* (Milan: Hoepli, 2002).

quotient. But of the inner pressure, we can only measure its gradient, since only the gradient of the pressure enters into the equation of motion: thus it is determined only up to a constant c, or possibly a function of time $c(t)$. But this restriction is nothing more than the price paid for an especially simple symmetry property of nonviscous fluids: the inner pressure used for describing the constitution of a fluid behaves like the potential of an especially symmetric force, for instance, of a central force. It denotes in fact the mechanical energy density.

From simple laws and theorems to entire theories

So far I have considered the role of concepts in only one law or one theorem. But what connects isolated theorems and laws? This is done, as everyone knows, by the most powerful connectors in science: by mathematical operations. It is worthwhile to look at this process with the help of a few examples. Beginning again with the gravitational force, we shall see how gravity became accessible to analytical mechanics.

Kepler had stated the results of his many observations in his three laws, probably the most popular ones ever discovered. While the first one is a purely geometric, the second and the third ones are kinematical statements. In his *Horologium Oscillatorium* Huygens refined Galileo's *principio*, formulated in the *Terza Giornata* (third day) of the *Discorsi*, and also sharpened Descartes's concept of momentum. Newton, to apply his concept of force, also had to propose explicitly the law which attracts the stars: the $1/r^2$ law of universal gravity. To deduce Kepler's laws he had to make heavy use of his new infinitesimal calculus.

In the *Principia* one sees that Newton solved the problem by concluding from the shape of the orbit, the areal law and the hypothesis of a central force $f(r)$ to the explicit form $1/r^2$ of this force. But he only partly proceeded and concluded in the inverse direction: given the law of the force: what is the shape of the orbit? Now more and more the great connectors took over.

In 1710, first Jacob Hermann and then Johann I Bernoulli solved the problem in the way we do it today, that is, by putting the concept of force into the central place and deriving all possible orbits from the given law of the force.[6] Bernoulli also cleared up the question of the number of arbitrary constants of integration, on which the solution depends as well. He also made it clear that for all central forces the calculation of all orbits could be reduced to one integration. But this is not all by far.

Starting from the concept of force and putting its explicit form into a differential equation made it possible to progress beyond the two-body problem. While so far for the three-body problem questionable assumptions had to be made, one could now at least write down the equations for any number of gravitating bodies, not only for two. Of course, it took time for all all this to be seen, but analytical mechanics was now on its way and, thanks to Clairaut, Euler, d'Alembert and many others, became the first branch of physics to be systematically organized by mathematics. The new concept, the force, which at the beginning did "no more," one might say, than adjoin mechanics to geometry, became now the centre of the new science, its organizer, as well as the key to the solution of innumerable new problems. For while concepts alone connect only few phenomena, together with mathematics they connect and organize an unlimited number of them.

---

[6] Jacob Hermann, "Extrait d'une lettre de M. Hermann a M. (Joh.) Bernoulli datée de Padoue le 12 juillet 1710," *Memoires de l'Ac Royale des Sciences* (Paris: Boudot, Paris, 1712); Johann I Bernoulli, "Extrait de la réponse de M. Bernoulli a M. Hermann, datée de Basle le 7 Octobre 1710," *Memoires de l'Ac Royale des Sciences* (Paris: Boudot, 1712).

This holds even more for the mechanics of continua and for electromagnetism. One reason is that classical mechanics is unable to provide the possibility of computing the structure of materials from scratch: quantum mechanics is needed. Classical mechanics must use "constitutive equations" as independent hypotheses, and then the vector and tensor calculus take care of the correct formulation. But while these relations, unless they are simply linear approximations, are at best good guesses, they are extremely useful.

There the interplay between the creation of new concepts – especially "constitutive concepts," such as, for example, "ferromagnet" – and the development of new mathematics adapted to these new problems, formed the basis of a fascinating drama, which began with Jacob Bernoulli.

Torricelli and Newton had laid the foundation. Galileo's *principio* was the dynamic principle. Following it as his guide, Daniel Bernoulli could, with his famous equation, unite hydrostatics and hydraulics, thus creating 'hydrodynamics', as he called it.[7] But in practice his equation was applicable only to flows through tubes and canals. Much more was needed to formulate a rigorous basis for the whole field.

It was Johann I Bernoulli who introduced Newton's force into this new domain, as Euler was quick to learn. D'Alembert gave an example of what is today called a field description: the formulation where the time and all space coordinates are used as independent variables. Of course, this powerful device called for a new mathematical calculus: partial differential equations. These were now studied more and more by d'Alembert and by Euler.

Euler had recognized the importance of "inertial systems," the only ones where the laws of nature possibly have such a simple form that one can hope that our imagination is capable of catching them. He had published this insight in 1750 in a paper, where he wished to use it for formulating a theory of rigid body dynamics. But the paper fell short of this goal; it would take Euler twenty-five more years to reach it in a completely satisfactory way.

However, he could apply it to hydrodynamics. What was missing there was precisely a new concept: the "inner pressure"! With this, everything fell into place; Euler could write down his four celebrated equations of hydrodynamics, together with a fifth one, the constitutive equation, which connects pressure and density according to the material in question (water, air, etc.). Thus hydrodynamics was the first field theory: a system of partial differential equations that are invariant under a group (here the Galilei group), which united, for instance, hydrodynamics and aerodynamics.

The fourth equation, the continuity equation, requires a special comment, as it leads to a conservation law for the mass. For this reason, as well as for its extension to the chemical processes by Lavoisier, the concept of mass is often equated with the concept of "matter." But the introduction of the mass, sketched above, shows that it has its place in dynamics, and Einstein's new dynamics says that this conservation law does not always hold. Matter, on the other hand, is not a scientific concept at all: it does not enter into any equation, nor into any prescription for performing an experiment. It is in fact a philosophic rather than a scientific concept, also used in other disciplines. Thus the doctrine of materialism is not supported by science at all.

There is no space here for treating the most complex and difficult domain in mechanics: the elastic bodies, rich in the most interesting dynamic and constitutive

---

[7] For Daniel Bernoulli's hydrodynamic studies see Clifford Truesdell, "Editor's Introduction: Rational Fluid Mechanics, 1687–1765," *Leonhardi Euleri Opera Omnia*, Series II, vol. 12, pp. VII–CXXV (Zurich: Orell Füssli, 1954).

concepts, but these can be studied in a wonderful book by Piero Villaggio, *Mathematical models for elastic structures!*[8] The third example is picked from the path to Wolfgang Pauli's exclusion principle. Very few activities before or since resemble the theoretical research done in the period between the Bohr-Sommerfeld theory of the hydrogen atom and Heisenberg's discovery of matrix mechanics, when theoreticians worked almost like experimentalists, in immediate contact with the experimental material. One main discovery that contributed to the progress of quantum mechanics was the Zeeman effect. If an atom emits light in a magnetic field, the lines will normally split into an odd number of lines, a fact that was well explained. However, sometimes the number of lines is even, and this "anomalous Zeeman effect" resisted all attempts to explain it.

Pauli was much preoccupied with this riddle. As he tells it,

> A colleague who met me strolling rather aimlessly in the beautiful streets of Copenhagen said to me in a friendly manner, "You look very unhappy"; whereupon I answered fiercely, "How can one look happy when he is thinking about the anomalous Zeeman effect?"[9]

Once I asked Pauli myself, how he had dealt with the spin when he discovered the exclusion principle; he said, *"Ich sprach von einer 'klassisch nicht beschreibbaren Zweideutigkeit'*," of a "classically non-describable two-valuedness." Looking for these words in Pauli's collected papers I searched for some time in vain. I found, for example, nothing in the article that contains his formulation of the exclusion principle. Eventually I came to one of his less famous papers, "Über den Einfluss der Geschwindigkeitsabhängigkeit der Elektronenmasse auf den Zeemaneffekt," which was received by the *Zeitschrift für Physik* on December 2, 1924.[10]

The aim of the paper was to disprove the conjecture, that the closed shells, especially the innermost one, the *Rumpf*, are responsible for the anomalous effect. Thus the argumentation is mostly critical. Only at the end does he offer his own proposition:

> According to this standpoint, the doublet structure of the alkali spectrum, as well as the violation of Larmor's theorem originate from a peculiar, classically non-describable two-valuedness of the quantum theoretical properties of the outer electron.[11]

A new concept? Certainly, even if seemingly a negative one. But the concept helped Pauli, at a time when no correct theory was yet mathematically formulated, to focus his attention in the right direction. Indeed, only six weeks later he could send his article, where the exclusion principle is formulated, to the *Zeitschrift für Physik!*[12]

---

[8] Piero Villaggio, *Mathematical models for elastic structures* (Cambridge: Cambridge University Press, 1997).

[9] Wolfgang Pauli, "Remarks on the History of the Exclusion Principle," *Science* **103** (1946): 213–215, p. 214.

[10] Wolfgang Pauli, "Über den Einfluss der Geschwindigkeitsabhangigkeit der Elektronenmasse auf den Zeemaneffekt," *Zeitschrift für Physik* **31** (1925): 373–385.

[11] *Die Dublettstruktur der Alkalispektren, sowie die Durchbrechung des Larmortheorems kommt gemäss diesem Standpunkt durch eine eigentümliche, klassisch nicht beschreibbare Art von Zweideutigkeit der quantentheoretischen Eigenschaften des Leuchtelektrons zustande* (Pauli, "Über den Einfluss...," *op. cit.*, p. 385).

[12] Wolfgang Pauli, "Über den Zusammenhang des Abschlusses der Elektronengruppen im Atom mit der Komplexstruktur der Spektren," *Zeitschrift für Physik* **31** (1925): 765–783.

But his mathematical formulation of the spin would only appear in 1927, after Schrödinger's discovery of the wave equation.

The consistency of a theory and the restrictions on its validity

How can we be certain that in a theory all theorems are logically connected? The best way to see this seems the Peano-Pieri-Hilbert axiomatic method, applied by Noll and Truesdell to the mechanics of continua. The idea is that if all statements are deduced from a few, the "axioms," which are consistent with each other, then the whole edifice is connected and consistent. I think that this growing unification is the true measure of the progress of the science of at least inorganic nature.

In my personal opinion, the teacher must also state, together with the axioms, the restrictions imposed on the theory. Such are, for example, the restrictive inequalities $a \ll b$, "a is much smaller than b." For instance classical, nonrelativistic physics holds only if:

$$v \ll c \text{ and } \int p dq \ll h.$$

These statements are of the same importance as the axioms. While the axioms assure the truth of a theory, working within these restrictions insures its validity. They prevent the scientist from overestimating the reach of what he learns and teaches. If the theory is applied outside them, it is simply not true!

Where will the unification of our theories lead us?

At the moment we consider inorganic science as organized and structured by four, or maybe five, interactions between the various fields and quantum systems, namely, the "strong," the "weak," possibly the "superweak," which interact between elementary particles and nuclei, the electromagnetic and the gravitational interactions. These last two are also accessible on the classical level, and after Einstein's formulation of general relativity, innumerable attempts, including several by Einstein himself, at unifying the two were made, but none succeeded.

But then in the frame of quantum mechanics a unification of the electromagnetic and the weak interactions centered around the four electroweak bosons, was at least partially successful.

It is the gravitational interaction that offers the strongest resistance. On the surface the reason is that Einstein's theory of the inertio-gravitational field is highly nonlinear, even in the absence of sources. For this reason this theory has not been formulated, so far, in the frame of quantum logic and quantum theory. But as I said, the inertia is a deep riddle.

Few, I guess, doubt that a unification of all five interactions will be realized one day. But what are the deeper obstacles? Which path will the process of unification follow? How long will it take? Even these questions cannot be answered today.

In such a situation one may be permitted to formulate a conjecture. Still today the sciences distinguish sharply between space-time on one side, and its content, the fields and quantum-systems, on the other. In my opinion this distinction, which is also taken for granted by most philosophers, will disappear one day. There will be theories where this distinction can no longer be made. But for formulating them, not only new dynamical concepts, but new prescriptions for measuring these quantities will be needed as well.

But there is another frontier of inorganic science that is perhaps even more tantalizing than the one mentioned so far. This is due to the discovery and experimental synthesis of larger and larger quantum systems, that is, molecules composed of a large number of atoms, which approaches the number of atoms of which some viruses are composed. But viruses are considered as "living," whatever that word may mean. Does our science arrive here at an insurmountable limit? And if not, can we progress with quantum theory as we know it today? Since there seems no need to appeal to Dirac quantum mechanics, the viruses would, in principle at least, be accessible to Heisenberg-Schrödinger quantum mechanics, or as perhaps we should say, quantum chemistry. The domain of quantum chemistry of complex nonstationary states is still very little explored, and with new mathematical tools this must not be impossible.

However, it is very likely that biology will also be mathematized. After all, we know already one mathematical law in biology: Mendel's. Certainly "mathematical models for biological structures" will be developed. With respect to them our present structures may then appear as degenerate limiting cases.

Conclusion

I hope to have presented here a small contribution towards making clear the role played by the concepts in the theories of nature. Why is this role so little perceived and underlined, especially in lectures?

One reason, I think, is the following. The development of mathematics is often stimulated by the natural sciences, but then mathematics is always based entirely on its own axiomatic foundation, independent of any applications. However, precisely this independence makes clear that its applicability to nature is a highly nontrivial fact!

Concepts, on the other hand, are introduced today unsystematically and ad hoc, in spite of many attempts, since Plato, to develop a systematic logic of concepts. Physicists often shrink from such an idea. Somehow they feel that it means that all laws of nature could then be deduced a priori. But this is not so. Rather, it means that theories could be presented with greater clarity. Moreover, we might then be able to find an answer to the perpetually haunting question: what do we really mean, when we speak of "the laws of nature"?

Acknowledgments

It is a pleasure to thank Prof. Stefano Bennati for the invitation to contribute to the volume in honor of Piero Villaggio, and my wife, Prof. Luigi A. Radicati di Brozolo, and Dr. R. Ziegler for discussions, suggestions and linguistic advice.

Introduction

The scientific ideas stated here and the few philosophical reflections, or to be more exact, the reflections made about philosophers, in this exposition, are intended as a contribution to the confrontation between sciences and philosophy. This confrontation has always seemed to me to be beneficial for both. Michel Ghins chose a topic where this confrontation is almost automatically present.[2] "Space and time" is a theme which, since the systematic discovery of geometry by the Greeks, has always incited philosophers and scientific to think about the questions that this discovery posed and still poses. In the beginning of his systematic exposition of the *Principia*, Newton dedicates quite a long chapter to the problems posed by space and time in his new mechanics.

It is necessary to note several things about Newton's doctrine:
- It is difficult, and sometimes even obscure;
- It is motivated by theological convictions;
- It was immediately the subject of an intense philosophical controversy;
- It was misunderstood or not even taken into consideration by most of the scientists who worked in the 200 years after Newton;
- The role that space and time play in mechanics and the problem that this role represents was correctly grasped by Newton.

Newton's mechanics probably had a deeper influence on philosophy than any other scientific discovery that followed it; we need only mention the names of Leibniz, Locke, Berkeley, Hume and Kant, to cite only the most well known. Even Kant, who had no doubts as to the necessity of a philosophical approach but who could not admit any theological foundations for scientific reasoning (contrary to Newton and Leibniz) recognized the new mechanics as a "touchstone" (*Probierstein*) for philosophy. Philosophy, according to Kant, is autonomous, but a contradiction between philosophy and science is necessarily the sign of an inherent error. According to Ernst Cassirer, it appears that Kant arrived to this view before anyone

---

[1] Given as a lecture on November 23, 1979 in the Sèminaire de Philosophie des Sciences and originally published as "Remarques sur l'espace et le temps chez Newton, Leibniz, Euler et dans la physique moderne," *Séminaire de Philosophie des Sciences 1979–1980* (Institut supérieur de philosophie, Université Catholique de Louvain) Rapport No. 15 (Louvain-la-Neuve: Cabay, 1980).
[2] Editor's note: Michel Ghins invited David Speiser to speak about this topic in the Sèminaire de Philosophie des Sciences. See also Michel Ghins, *L'inertie et l'espace-temps absolu de Newton et Einstein, une analyse philosophique.* Académie Royale de Belgique, Classe des lettres, 2nd series, vol. 69 (Brussels: Académie Royale de Belgique, 1990).

else thanks to several of Euler's writings which he had studied and which I will mention in what follows.

This is already reason enough why Euler's writings on space and time deserve to be read, but in addition, they constitute the first modern investigations, that is, those written entirely in scientific language, and for a long time they provided the clearest exposition of this problem. Today, the notion of group makes it much simpler to formulate the doctrine of space-time, especially the "doctrine" according to Einstein and Minkowski.

If mechanics constitutes a "touchstone" for philosophy, it is because its birth marked an irreversible step. But a step in what direction? And what is the significance of this step? Personally, I'm not one of those who think that philosophy was made superfluous by the birth of mechanics, and that mechanics alone provides all the valid answers. It is sufficient to pose a question about the scientific method itself, as every scientist does constantly: whatever the answer may be, by definition it is not scientific, but rather philosophical. But I do share the viewpoint of those who think that a new situation was created, and my answer would simply be the following.

We don't know exactly how modern physics advances, nor why it achieves success, except that obviously "we hit on something"[3] (it is difficult to express this in the language of Descartes), and our common understanding is increased by a giant step by physics, chemistry and molecular biology; these sciences constitute, we might say, an immense "coordinate system" for other sciences. Certainly, scientists are far from agreeing on all questions. They don't even always agree on their foundations. However, there is a common agreement regarding their methods, which makes it possible (at least in the inorganic domain) to situate the problems so that one can work on the scientific problems jointly (and also, by the way, jointly make money). What irritates me in the method of philosophy is that here, contrary to what happens in science, one finds with each philosopher a new start from the beginning, and an explicit refusal – or at least an implicit neglect – of all the results acquired by most earlier philosophies.

Is it clear that this is not without rhyme or reason, but must it indeed be so?

The encounter of Newton's mechanics with the different philosophies that followed seems to me to be an example that always poses questions that are still current:

– What is there in science that is important for philosophy?
– What does the scientist working in his discipline expect from a philosophy, in order to come into contact with it?
– What does the scientific method mean for the philosopher?

In what follows, I restrict myself to the problems posed by space and time and to discussing how they are conceived by the physics of yesterday and that of today.

Newton's doctrine of space and time, and the questions it poses to the philosophers

What was the situation created by Newton's mechanics?

What questions does this mechanics pose to the philosopher?

Why an absolute space and an absolute time? The two, it is true, might appear as erratic rocks in a landscape to which they don't naturally belong.

---

[3] In English in the original text.

The answer to the first question can be given briefly with two examples. Starting from a reduced number of equations Newton's mechanics makes it possible to calculate – that is, to predict – all phenomena related to the propagation of a system in space and time. That is to say, starting from some general axioms and a constitutive principle (the law of gravitation) the astronomer can (in principle) predict, for all ulterior instants and with all the desired accuracy, the position and the speed of every planet or moon, provided that these magnitudes are known at one instant which is, moreover, arbitrary. Or, using the laws of Ampère and Faraday as well as the numeric results of some measurements, an engineer can construct an electrical circuit such that the user only has to flip a switch to have light in his office.

We can see from these two examples, one borrowed from science, the other from technology, that theory (that is, a system of laws formulated in a mathematical language), observation and experience are closely related. This remark brings us closer to the second question: what are the questions posed to the philosopher by the new science? At the heart of this problem one finds the enigmatic relationship between mathematics and the world of the notions and logical concepts on one hand, and the world of intuitions and observations on the other. I say "enigmatic relationship" because these two worlds are separated by an abyss, and yet experiment manifestly establishes a tie between the two. The foundations of mechanics must therefore contain, in addition to the equations, <u>the prescriptions that tell us how these equations must be applied to phenomena</u>, or if you will, how certain observations – for example, the angles under which we observe the planets – enter in the equations. What might such prescriptions be? For the moment we don't have an answer to this question, but there is certainly nothing in either so-called pure mathematics or in the phenomena themselves that teaches it to us. As the Greeks had already shown, there is an abyss between these two worlds. It is here where the doctrine of absolute space and time comes in, because it tried to put the two worlds in contact. This doctrine makes it so that the equations can be applied by means of additional restrictions that must be respected if one wants to achieve a correct result. If this doctrine of Newton's seems obscure to us, it is because its author, like his successors for a long time to come, didn't possess the necessary mathematical language to formulate it clearly: group theory. For the simplest systems, points with mass, the laws of Newton amount to the following statements (here I am not following Newton's order):

- The propagation of a system is determined by the equilibrium between the internal forces and external forces. External forces are related to observable systems (e.g., the sun, the moon) which are then, incorrectly, called "causes," since they seem "to cause" some accelerations. In contrast, internal forces are not related to observable causes; they are only related to space-time itself, which thus acquires a physical dimension;
  In other words:
    - accelerations have an absolute significance that is due to the internal forces ("to inertia");
    - positions, velocity[4] and momentum in contrast have only a relative significance, that is, these magnitudes are subject to the principle of relativity, which says that space-time is a homogeneous <u>linear manifold</u> on which acts (and on which only acts) the Galilei group.

This is obviously a modern formulation, which Newton could not use. Moreover, the Galilei group is, alas, a complicated and difficult mathematical structure to

---

[4] Editor's note: velocity refers to the vector quantity, speed is the magnitude; in French they have only one word for it, *vitesse*. I thank Frans Cerulus for pointing this out to me.

translate into simple words. We cannot therefore be angry at Newton for having been obscure, nor even for having fallen into error (later I will use Einstein's mechanics as an example, and I will speak of Minkowski's space-time on which acts the Poincaré group, whose structure is more transparent). This statement expresses the inertia or inertial force due to space-time.

- These two statements are related by a third that says there exist frames of reference in which the laws of Newton's mechanics are valid. These so-called inertial frames of reference are all related by the Galilei group. In contrast, in relation to all other frames of reference, the laws of mechanics are not valid in the simple form that Newton found, but in a form that must be explicitly calculated while taking into account the transformation that relates the frame of reference used to an inertial frame of reference. But in such a system, it is generally no longer possible to distinguish the internal forces (inertial) from the external forces.

There thus exists, imposed by nature if you will, a class of privileged frames of reference in relation to these laws: we might say that this whole class constitutes absolute space-time.

Note that for the following three statements:

1. there exist absolute frames of reference
2. these frames of reference are related by a group that acts on a linear manifold
3. this group is the Galilei group

the theory often called "special relativity" only differs in relation to the third, while it remains in agreement with the first two. (Einstein wanted to tackle the first, however, in my opinion he has instead modified the second; I come back to this.)

Newton didn't possess the notion of group; it was therefore difficult for him to express that mechanics conceives an infinity of Euclidean spaces that are interpenetrating and mutually in relative motion, whereas the mathematical ideas seem to require a unique Euclidean space. Moreover, he himself didn't entirely grasp his own idea. In Book III of the *Principia*[5] he says that the center of inertia of the planetary system is at rest. This affirmation doesn't make sense in the context of his mechanics. But what matters, it is that with absolute space and time Newton was able to bridge the abyss between the world of mathematics and that of phenomena observed (by our senses). This done, performing an experiment becomes a method that makes it possible to provide a reproducible result. And this result is moreover still a number that expresses a relation between objects (observed objects and "devices" of measurement).

Leibniz's critique

It is certainly worthwhile to discuss the viewpoint and critique of Leibniz. Leibniz was not only a great mathematician (one of the greatest), to whom we owe the independent discovery of infinitesimal calculus (and in a more modern form than that in which Newton had presented it), he had also worked successfully in physics. In this regard it is interesting to read the assessment of Clifford A. Truesdell, who wrote that Leibniz's achievements in the theory of elasticity are sufficient to place him among the best; he especially emphasized Leibniz's feeling for experiment.[6]

---

[5] Isaac Newton, *Principia*, Book III, Prop. XI, Theor. XI.
[6] See Clifford A. Truesdell, *The Rational Mechanics of Flexible or Elastic Bodies, 1638–1788, Leonhardi Euleri Opera Omnia*, Series II, vol. 11b (Zurich: Orell Füssli, 1960), p. 128.

One of the greatest contributions to our understanding of nature was to have grasped the importance of the notion of <u>function</u>, notably of analytic function, as we call it (at the time it was called *functio continua* because these distinctions were not yet grasped). This notion of function is fundamental in physics, that is, for any mathematical understanding of nature; it is much more important, for example, than the often inadequate notions of matter and cause that are often improperly used and lead to an incorrect comprehension.

In what follows, I will try to give a layman's (that is, a non-specialist's) response to the following question: Why is Leibniz's philosophy still alive, or if you prefer, why does it still pose interesting questions? It seems me that the answer must be sought in the following direction. Leibniz was one of the philosophers who had profoundly understood the role that mathematics plays in the scientific understanding of nature. This means, among other things:

1. Leibniz, by dint of his own discoveries, had acquired a profound understanding and broad vision of what mathematics is. Through his discovery of infinitesimal calculus, following a study of a work by Pascal,[7] he came to understand that *esprit de finesse*[8] was not opposed to mathematics, but that very much the contrary was true.

2. Leibniz was among the first mathematicians of the modern age not only to have sensed the beauty of mathematics (all mathematicians worthy of the name had always sensed that), but to have also understood that this beauty itself must be represented by writing which has to reveal the deep and hidden structures. These structures and orders lead us to notions like, for example, the "determinant" discovered by Leibniz, which in a single stroke makes a whole domain – in our case, that of linear equations – transparent.

3. The logician is among the first to address the foundations of mathematics, and see that its role in the sciences is not reduced to constructing a formal tool which is then introduced into a science that already exists without it, to make statements concerning matter (whatever the term "matter" might mean in this case) quantitatively exact. Leibniz saw that without mathematics it is impossible to speak of nature.

Leibniz was, it seems to me, the last philosopher who proposed a system which takes into account the nature of mathematics, its power of invention (*methodus inveniendi*), and the role that it plays and must play in science.

I have never been given a satisfactory answer to the question that is central for an understanding of Leibniz's philosophy: what is a monad? What was Leibniz thinking about? If while reading what he wrote I try to find something that gives an account of its description, it is what the mathematician calls an "element of analytic function" (in the sense of Weierstrass), which seems me to come the closest. (In fact, it is amusing to see, as always, that from time to time, a physicist takes a series of Leibniz's ideas that group themselves around the idea of analyticity in support of his

---

[7] Editor's note: The "Traité des sinus du quart de cercle" (1659) was published in the letters of Amos Dettonville, a pseudonym of Blaise Pascal. In *Œuvres de Blaise Pascal* (Paris: Chez Lefèvre, 1819), vol. 5, p. 312–325.

[8] Editor's note: an expression coined by Pascal in his essay on the "Esprit de géométrie," first published posthumously and incompletely in 1728. The *esprit de géométrie* allows the mathematician to solve a problem of nature once it has been translated in mathematical language, working within a given set of axioms and rules. On the other hand, a man ignorant of mathematical technique, may be able to discover and discern very finely the multiple and often hidden causes of a phenomenon and draw the right conclusions by intuition: he is gifted with *esprit de finesse*.

program. The last of these programs came about, only a few years ago, in California.[9])

We have to wonder then why Leibniz didn't grasp the essential content of Newton's doctrine on space and time, because this doctrine is indispensable in order for mechanics to provide correct results, even though it can definitely be formulated in a better way than its author did. But the implementation of classic (and quantum) physics leaves no doubt today, and thus it is clear that Newton was right regarding the essence of this question. I hope to show that likewise in the general theory of relativity the situation is not the one proposed by Leibniz. But, on the other hand, it is quite true that, from the philosopher's point of view if you will, the critique of Leibniz marks, as I said, a point, and it is this which first of all makes the "Leibniz problem" interesting and topical. It also explains why, following the discovery of the general theory of relativity, his name turned up again in the discussions.

But there is more. In reading Leibniz's texts (of which I unfortunately know only a few), I get the impression that the cosmos of Leibniz is on the one hand philosophical-mathematical and, on the other, historical, that is to say, a succession of unique events. But, between the two, nature and the science that explores it seem to be missing. I might be mistaken, but something, some element indispensable to the scientific method, seems to be missing. But what? I shall return to this later.

Euler's investigations of space and time

It comes as no surprise that we find in Euler the first great steps towards the understanding of this set of problems; Euler was in any case the first to approach the notion of group. One of his first approaches is found precisely in a work comparing wave optics to corpuscular optics.[10] He shows that the difference between the two is due to the fact that the corpuscular optics automatically admits the principle of relativity.

It is also not surprising to see Euler insisting on the transcendental character (as Kant said) of space and time, that is, that they exist outside of us but that they are also entirely accessible to our reasonings, i.e., they are also in us (see the *Letters to a German Princess*).

Apart from a few scattered remarks, Euler expressed himself at least five times in a systematic way regarding the nature of space and time, on the problem of inertia, and on the role that the three play in the foundations of Newton's mechanics. These are:

1. The *Mechanica sive motus scientia analytice exposita* (E15 and E16) of the year 1736;[11]
2. A relatively brief article, *"Réflexions sur l'espace et sur le temps"* (E149) published in the *Mémoires de l'Académie de Berlin* in 1748;[12]

---

[9] Editor's note: this refers to the so-called S-matrix theory initiated by Geoffrey Chew which advocates a theory where self-consistency acts as a Leibnizian sufficient reason and which relies heavily on the theory of analytic functions.

[10] Leonhard Euler, "Explicatio phaenomenorum quae a motu lucis successivo oriuntur" (E127), *Commentarii academiae Scientiarum Petropolitanae* 11 (1739), pp. 150–193. Reprinted in *Leonhardi Euleri Opera Omnia*, Series III, vol. 5, pp. 47–80 (Basel: Birkhäuser).

[11] Leonhard Euler, *Mechanica sive motus scientia analitice exposita* (E15 and E16) (St. Petersburg: Ex. typographia Academiae Scientiarium, 1736). Rpt. *Leonhardi Euleri Opera Omnia*, Series II, vols. 1–2 (Zurich: Orell-Füssli).

[12] Leonhard Euler, "Réflexions sur l'espace et le temps" (E149), *Mémoires de l'académie des sciences de Berlin* 4, (1748) 1750, pp. 324–333. Rpt. *Leonhardi Euleri Opera Omnia*, Series III, vol. 2, pp. 376–383. (Basel: Birkhäuser).

3. The treatise entitled *Anleitung zur Naturlehre* (E842).[13] Published in 1862, many years after Euler's death and incomplete, since six pages were missing from the manuscript, we know that the *Anleitung* was written after 1745, but nothing more. Hermann Weyl said of this work, "in magnificent clarity [Euler] summarizes the foundations of the philosophy of nature of his time";[14]

4. His magnificent *Theoria motus corporum solidorum seu rigidorum* (E289) of the year 1765.[15] It is here that we find his most mature ideas;

5. The *Lettres à une Princesse d'Allemagne* (*Letters to a German Princess*, E343) of 1772.[16]

A systematic review of his works could bring to light other important passages.

However, before speaking of what is contained in these works, let us make a remark. Their position at the beginning of a treatise on mechanics or of an exposition of the principles of natural sciences clearly shows that for Euler space and time – that is, the space and time we live in – are the object of the science of nature, and therefore of physics. There is therefore no geometry that is dissociated from physics and that precedes it, as in, for example, Kant or Poincaré. Geometry is a physics, if you will, as it had been for the Greeks and for Newton, and would later be for Riemann and Einstein. The isolated article *Réflexions* of 1748 confirms this observation: it is the role played by space and time in mechanics that teaches us their true nature.

Now let us say a few words about these works.

The *Mechanica* of 1736 is not only the first book where Euler addresses the problem, it is the first original and systematic setting forth of Newton's mechanics published after the *Principia*. Jacob Herrmann's *Phoronomia* did not produce much original material, and the mechanics of Johann I Bernoulli was not printed. It was therefore from this book of Euler's that scientists most often learned the new science. But since later presentations will be more transparent, I won't stop at the *Mechanica*.

The *Réflexions sur l'espace et sur le temps* has become the best known work that Euler dedicated to this question. This is due to the fact that they found an illustrious reader for whom Euler's theory was decisive: Kant. It appears that the Austrian philosopher Alois Riehl was the first to have drawn anew the attention of historians to this connection. He cites the following instances in which Kant speaks of Euler:[17]

---

[13] Leonhard Euler, *Anleitung zur Naturlehre* (E842). In *Opera postuma mathematica et physica anno MDCCCXLIV detecta*, vol. II, pp 449–560 (Petropoli: Egers; Leipzig: Voss, 1862). Rpt. *Leonhardi Euleri Opera Omnia*, Series III, vol. 1, pp. 16–180 (Basel: Birkhäuser).

[14] Hermann Weyl, *Philosophy of Mathematics and Natural Science* (Princeton: Princeton University Press, 1949), p. 42.

[15] Leonhard Euler, *Theoria Motus Corporum solidorum seu rigidorum...* (E289) (Rostochii et Gryphiswaldiae : litteris et impensis A. F. Röse, 1765). Rpt. *Leonhardi Euleri Opera Omnia*, Series II, vol. 3 (Zurich: Orell-Füssli).

[16] Leonhard Euler, *Lettres à une Princesse d'Allemagne sur divers sujets de physique & de philosophie* (E343), 3 vols. (St. Petersburg : De l'imprimerie de l'Academie Imperiale des Sciences, 1768–72). Rpt. *Leonhardi Euleri Opera Omnia*, Series III, vol. 11 (Basel: Birkhäuser).

[17] Alois Riehl, *Geschichte und Methode des philosophischen Kritizismus,*" Vol. 1 of *Der philosophische Kritizismus und seine Bedeutung für die positive Wissenschaft. Geschichte und System*, 3 vols. (Leipzig, 1876–1887); see 2nd ed., 1908, p. 334 ff.

- in *Versuch den Begriff der negativen Grössen in die Weltweisheit einzuführen* (Attempt to Introduce the Concept of Negative Magnitudes into Philosophy, 1763);[18]
- in *Von dem ersten Grunde des Unterschiedes der Gegenden im Raume* (On the First Ground of the Distinction of Regions in Space, 1768);[19]
- in *De mundi sensibilis atque intelligibilis forma et principiis* (Dissertation on the Form and Principles of the Sensible and the Intelligible World, 1770).[20]

But the credit for having brought to light the question around which the discussion revolved and for having determined what Euler's important contribution was belongs, in my opinion, to the Neo-Kantian Ernst Cassirer, who dedicated to this one chapter in his book *Kants Leben und Lehre*[21] and another in his *Erkenntnisproblem in der Philosophie und Wissenschaft der neueren Zeit.*[22]

Cassirer's 1921 book on the theory of relativity begins with a quote from Kant from the *Attempt to Introduce the Concept of Negative Magnitudes into Philosophy* in which he mentions Euler. Kant wrote:

> The celebrated Euler, among others has given some opportunity for [keeping the metaphysical speculations on time in the track of truth], but it seems more comfortable to dwell in obscure abstractions, which are hard to test, than to enter into connection with a science which possesses only intelligible and obvious insights.[23]

But *Réflexions* has the character of a sketch. The problems are mentioned and indicated and the difficulties are shown, but without being pursued systematically. *Réflexions* doesn't possess the organization and clarity of detail that are the mark of Euler and that we will find in the last two works that concern us here, the *Anleitung* and the *Theoria motus*. Euler's goal in the *Anleitung* is to provide a systematic and philosophical view, if you will, of the whole science of inorganic nature. He tried to reduce his statements and equations to a small number of principles and properties (*Eigenschaft*, as he says, but this expression is unfortunate). It deals with:

- extension;
- variability;
- durability;
- impenetrability.

One can easily see the connection of these properties to:

- space;
- time;
- inertia;
- forces.

[18] In *Immanuel Kants Werke*, Ernst Cassirer et al., eds., 11 vols. (Berlin: Bruno Cassirer, 1912–1923), vol. II, p. 206.
[19] *Ibid*, p. 394.
[20] *Ibid*, p. 431, 436.
[21] Ernst Cassirer, *Kants Leben und Lehre* (Berlin : Bruno Cassirer Verlag, 1921).
[22] Ernst Cassirer, *Das Erkenntnisproblem in der Philosophie und Wissenschaft der neueren Zeit*, 4 vols. (Berlin: Bruno Cassirer Verlag, 1906–50).
[23] Ernst Cassirer, *Substance and function & Einstein's Theory of Relativity*, Supplement: Einstein's Theory of Relativity Considered from the Epistemological Standpoint (Open Court Publishing Company, 1923; rpt. 1953 & 2003, New York: Dover), pp. 351–352.

To appreciate the connection between impenetrability and force, it is necessary to remember that Euler worked in the domain of continuum mechanics. He had discovered, among other notions, that of "dynamic pressure," which is today the basis of hydrodynamics and which is situated in the line that will lead to the discoveries of Cauchy and Maxwell, to the mechanics of Kirchhoff and to those of Heinrich Hertz, a line on which are also located certain current approaches concerning field theory. We don't know why Euler didn't publish the *Anleitung*, perhaps simply because he didn't find a publisher. This is indeed what nearly happened to the *Theoria motus*, to which I will return in a moment. Let's note further that Cassirer doesn't seem to have known the *Anleitung*; he doesn't mention it in either his *Erkenntnisproblem* or in *Kants Leben und Lehre*.

On the other hand, I am indebted to Cassirer for having drawn my attention to the *Theoria motus*, which represents a decisive step beyond everything that had been said before, in particular in the *Mechanica* and the *Réflexions*.

Here is a brief summary of the first two chapters of the *Theoria motus* (I differ slightly from Cassirer):

Euler's method is two-fold: in the first chapter, he introduces kinematics (as we would say); in the second, dynamics. The two chapters are quite long (some 60 pages), but the main theses of kinematics are:

- motion (not only velocity) is relative and must be determined in relation to at least four points or in relation to a body. This means that velocity, acceleration, etc. are relative. Euler means to say by this that a measurement (as for example in astronomy) is made always in relation to a system; Therefore, motion and rest are indiscernible: *cadit ergo celebris illa distinctio...*;[24]
- on the other hand, it doesn't mean that space and time are imaginary magnitudes or only invented. Time must flow <u>independently of us</u>, otherwise we could not take measurements, that is, we would not have any watches. (Analogous remarks are valid for space.)

If absolute space was rejected in kinematics, the picture now changes when we examine the principles of dynamics. Chapter two is dedicated to an examination of what he calls <u>internal principles</u> of dynamics, thus, as he will say, to inertia. Note that he says, among others things, that if one wants to compress a body, forces are needed: therefore, as far as dynamics is concerned, the size of a body is something absolute.

What does mechanics say? The answer is:

- In effect, rest and uniform rectilinear motion have only a relative significance.
- In fact, it doesn't matter whether a body in relation to an observer is at rest or in uniform rectilinear motion; time and uniform velocity are merely <u>a state</u> of the body that is maintained in absence of forces. (It may be that here we find the first explicit <u>calculation</u> that shows invariance under a "Galilean transformation." The notion of *status* is of course found in Newton, but presently the last trace of a distinction between rest and uniform and straight velocity disappears).
- One <u>can</u> therefore (contrary to the followers of Leibniz) consider a single body in the world, it merely means that the body is isolated.
- But the essential point is that the change of a state – that is, the acceleration – has an <u>absolute</u> significance. Further, if one wanted to define this acceleration

---

[24] "Hence the celebrated distinction [between motion and rest] falls...," Euler, *Theoria Motus op. cit.*, vol. 1, ch. 1, p. 8 (see *Leonhardi Euleri Opera Omnia*, Series II, vol 3, p. 24.

only in relation to other bodies, one would fall into an infinite regression (later, Euler will also show that some systems – the inertial systems – are distinguished).

- If one does not accept such an absolute space-time, mechanics cannot function, i.e., it cannot be applied.
- Also note that, from a historic viewpoint, apart from the infinitesimal calculus of Leibniz, here we find explicitly the affirmation that rest and uniform velocity are dynamically indiscernible (in Newton this idea was not yet entirely found). On the other hand, it doesn't matter much if Euler's axiomatics do not entirely coincide with those of Newton: they agreed on the principles.

(The *Theoria motus* is one of the first where the Cartesian systems are used systematically in mechanics.)

If Euler's explanations seem longer than necessary, this is because, as I said earlier, the fundamental notion of <u>transformation group</u> had not even been conceived yet. Without this powerful notion, it is difficult to formulate the bases of mechanics clearly.

On the other hand, there is no doubt that discussions with philosophers had sometimes driven Euler to give too much credit to the mere words.

In any case, we maintain that the problem is seen and analyzed much more clearly than in the great majority of the texts that followed it over the course of the next 150 years.

Minkowski's space-time (Einstein's mechanics of 1905).

What does physics teach us about space and time? To simplify things, I will give the answers brought to us by Einstein's mechanics of 1905 (the theory of special or linear relativity). These answers don't differ fundamentally from those of Newton's mechanics, but they are more transparent; it is interesting to note, and it is necessary to point out, that Newton's famous first axiom can be formulated much more simply in Minkowski's space-time.[25] It is sufficient to say that the system moves on a straight line in space-time, and it is superfluous to add at "uniform speed." The principle of relativity now says that the laws of nature are invariant under the group of space-time isometries.

I will return later to the general theory of relativity (Einstein 1908-16) but I am not going to discuss quantum mechanics at all.

Einstein's mechanics makes four statements about space-time:

1. Space and time form a four-dimensional manifold;
2. This manifold is a linear space;
3. From each point is issued the light cone on which light propagates itself in the vacuum. This regardless of the state of the source. (The light cones govern what one might call causality.);
4. The distance between two points of space-time (between two "events") has meaning as a measurable quantity, regardless of the observer's state in space-time (from which it follows, for example, that if two observers are at rest in relation to each other, they can compare their measurements of

---

[25] See Hermann Minkowski, "Das Relativitätsprinzip," lecture delivered before the Mathematische Gesellschaft in Göttingen, November 5, 1907, published in *Jahresbericht der deutschen Mathematiker-Vereinigung* 24 (1915): 372–382 and in the *Annalen der Physik* 352, 15 (1915): 927–938.

length and time directly). (Eintein-Minkowski space-time especially differs from that of Newton-Euler in point 3, the existence of the light cone, an absolute that separates past, present and future.).

It is important to note that it is light – that is, the electromagnetic field, and therefore a non-geometric agent – and not kinematics that defines these limits as well as that which is sometimes called causality. We arrived at this kinematic knowledge through the laws of electrodynamics. (In Newton's mechanics, analogous limits are defined by ideal rigid bodies that permit the transmission of instantaneous signals; the light cone is then entirely open to form an absolute present.) Otherwise, the two mechanics are essentially in agreement. They are particularly in agreement on the existence of straight lines and on their distinct dynamical status, affirmed by the first axiom of Newton, which says that in the absence of an external force, the body moves along a straight line (it is this point and this point alone that will be modified by general relativity).

The axiom cited says "in the absence of external forces," but doesn't add "in the absence of other bodies …;" bodies don't play any role and their presence or their dynamic state are not taken into account at all.

Euler expresses this by saying that the motion of the body is set with respect to space and time and not in relation to other bodies. He illustrates this by an example taken from hydrodynamics: a boat at rest in water (Euler says in "still" water). If the water begins to flow, the boat little by little also begins to move, but this acceleration is not governed by the geometric relations between the boat and the different parts of water. But, because this example is complicated, I have chosen another, simpler one. Take a rotating wheel on the rim of which is located a small body.

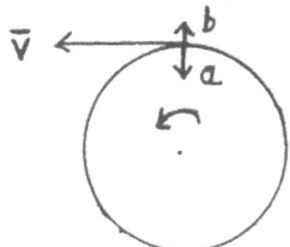

In the system defined by the wheel, the small body is held by the rim, which exerts a centripetal force (*a*) on the body. But this is counterbalanced (action-reaction) by a centrifugal force (*b*), without which the body would move toward the center. Where does this force come from? It is the effect of inertia (*vis inertiae*), a property inherent to empty space-time, which, however, distinguishes the straight line from other curves. Because, once the rim is removed, the centripetal force disappears and with it the centrifugal force. The small body leaves the wheel to continue on its way along a straight line of space-time, that is, at constant velocity. This is independent of the properties of the wheel (whether large or small, in wood or gold, heavy or light, whether it continues to turn or not, etc.). One can even remove it entirely without in any way affecting the fact that the small body continues to propagate itself along the straight line. This is what Newton and Euler had called *vis inertiae* or inertia. But one can still wonder: Does this inertia have its origin in very massive distant bodies, and therefore in the fixed stars? I believe it was Mach who posed this question, since he could not admit that an agent that determines the motion of a body comes from a source (from a "cause," if you will) that is not identifiable <u>by our senses</u>. It is evident that the so-called positivist view, which Mach held, and following Hume, would no longer be valid if the propagation of a system were even partially determined by something that would not directly affect our senses (e.g., "inertia").

Be that as it may, Euler posed (or in the case that it had been asked by another before him, repeated) the question as early as 1748. At that time, it was difficult to him to answer in the negative. How could we in fact exclude such a proposition by observation alone? The answer is, of course, by showing that the motion of a body changes with the position of the fixed stars. However, at that time, no motion of the fixed stars had been ascertained. The first parallax would only be measured a century later; the motions of the stars were only determined during the twentieth century. Euler therefore had to content himself with:

- noting the absence of all theory that would have formulated such an idea mathematically;
- saying that thus nothing permits us to think that if a body (a satellite) passes close to a fixed star that, leaving gravitation aside, it "regulated its motion according to this star."

This is the answer given by the mechanics of Newton as well as those of Einstein.

It should be underlined again that the existence of the light cone is postulated a priori and does not depend on the physical state of the systems in space-time (we will see that, according to the general theory of relativity, the situation differs somewhat but not essentially). It would be absurd to say that it is for dynamic reasons that space-time has four dimensions. Thus, space-time possesses the physical properties that must be taken into account when one wants to determine the propagation of system.

The present situation: general relativity, quantum mechanics and dynamics of elementary particles

We must say a word about the present situation of our knowledge of space and time. The physics of the twentieth century provided us with a concrete and solid confirmation and posed two forbidding (or perhaps rather promising) questions.

Let's first consider the confirmation provided by the general theory of relativity. Of the four characteristics enumerated earlier (dimension, linearity, light cone, and conformal nature of the metric), this time it is the second point that will be modified. Space becomes a Riemann space, that is, a space with variable and dynamic metric, a space that, as Riemann had already predicted, can be influenced by its material content. According to Einstein this is represented by the energy-momentum tensor. Therefore space-time had now become <u>manifestly</u> a physical agent. But, on the other hand, it preserves the properties that physics doesn't affect: dimension, the light cone (locally), and (contrary to what Einstein had hoped) the local metric character (i.e., the Riemannian <u>shape</u>) are given a priori. We can speak therefore of a geodesic and construct locally the privileged frame of reference, which allows us to distinguish the inertial-gravitational field from all other fields. It is from this point of view that, in my opinion, "Mach's principle" must be seen. It is true that the stars contribute to determine the actual inertial-gravitational field (as do all terrestrial objects, ourselves included). But it doesn't exclude the fact that space-time possesses invariant characteristics which dynamics, according to our present understanding (that is, according to the present theory) <u>doesn't deal with</u>, and which <u>are not directly accessible to our senses</u>. These are:

- dimension
- the light cone
- affine connection
- the quadratic nature of the metric.

This is related to the fact that the non-excited field (contrary to what happens in electrodynamics) is not a null field, which would be a singular space-time, but the empty space-time of Minkowski! And, with the exception of the situation of black holes, excitations due to the observable sources are always minimal (that is, the deviations from the geodesics in the propagation of a system as well as the components of the Riemann tensor are normally small because

$$\frac{GM}{c^2 R} \ll 1,$$

where R is a characteristic length of the system).

The two questions are posed respectively by quantum mechanics and the theory of elementary particles.

Quantum mechanics taught us that the laws of nature have a different logical structure than the one that holds in classic mechanics. The key notions are: "principle of superposition" and "modular logic."

Electrodynamics has been reformulated in accordance with these principles and the strong and weak nuclear forces could never have been formulated otherwise. However, up to now gravitation, and along with it, inertia, have resisted such an adaptation, or "quantification" as we say. It is clear that this "quantification" may have some surprises in store for us, surprises which might force us to reconsider seriously yet again our ideas about space and time.

It is the same with the attempt to formulate the physics of elementary particles. Important progress has been achieved in recent years, but we are still up against a wall as far as the "mass problem" is concerned (that is, the questions of why only certain values of mass are realized in nature, and of the role played by mass). However, we experience inertia as much through mass as through energy-momentum. What the physics of elementary particles has taught us up to now is that, apart from space and time, there exist degrees of freedom called (unfortunately) "internal" which must be taken into consideration. Thus completely new perspectives have been opened.

Conclusion

Where will we be led by the discoveries of the future? We learned from Newton's discoveries, and even more from those of Einstein, that in nature, one would not know how to speak of space without <u>also</u> speaking of time <u>and</u> inertia <u>and</u> forces. From the point of view of physics, geometric space is only one aspect, an abstraction if you will.

All that we can say for the moment is that space and time will certainly be still more intimately linked to the other realities of physics (for example, the electromagnetic field, the strong interactions, the internal symmetries of elementary particles) and that some of these new realities are even more elusive to the direct perception by the senses.

I would like to come back to Leibniz. It is clear he didn't grasp a fundamental point of Newton's physics. But it seems to me that it is necessary to say more. As I mentioned, it seems that he missed an important element in his philosophy. But what is it? Leibniz was able to discern the importance of the principle of "sufficient reason," but he seems to have applied it badly. This principle is intimately related to the idea of symmetry still discussed in physics today, perhaps more than ever. How does it come into physics? The answer is, in ascertaining a <u>homogeneous</u> essence (a manifold) according to which the facts are observed. In our case, space-time. It is this homogeneity that acts as an indispensable key for the foundation of classic <u>and</u> quantum physics. And quantum physics takes a second step in this direction. One often speaks here of homogeneous matter. In classic physics, this expression doesn't mean much, but in quantum mechanics it is different. It finds its strict expression in the indiscernibility of the elementary particles, which has calculable consequences that can be observed and thus verified experimentally. Leibniz didn't see the importance of the idea of a homogeneous essence for science. Earlier I said that in Leibniz there are mathematics on one hand and history on the other, and that he doesn't really seem to have a place for natural science. It may be that, accepting the vocabulary of Kant, one could say that he knows neither the <u>problematic</u> judgments nor the category of the <u>possible</u>. It is remarkable that the <u>possible</u> plays a much more important role in quantum mechanics, and I wonder whether this notion should not be entirely reconsidered.

Translated from the French by Kim Williams

Acknowledgments

I thank Patricia Radelet-de Grave and Michel Ghins for some bibliographic references, and Mr. Ghins for his interesting remarks.

Gruppentheorie und Quantenmechanik: The Book and its Position in Weyl's Work[1]

Hier sollen alle diese Fäden zu einem einheitlichen Ganzen verwoben werden.

H. Weyl

One of the remarkable facts in the history of Quantum Mechanics is that after Heisenberg's breakthrough in 1926, most of the important discoveries that were to follow occurred during the next few years. Many of them are connected with Schrödinger's discovery of wave mechanics. How was this explosion possible? Of course, on the *physical side*, the ground was prepared by work that had been done since 1900. This is a well known story and need not be repeated here. In an obituary on Wolfgang Pauli, Carl Friedrich von Weizsäcker wrote, "In his article about quantum theory in the first edition of the Blue Handbook (1925) he accomplished the work of art of answering practically every physical question correctly right before the discovery of the quantum mechanics, and therefore with seemingly still inadequate aids."[2] But it is much less well known that on the *mathematical side* much the same had happened, and only a wrongly placed pride in what they believe to be the autarchy of their own field can prevent physicists from noticing and learning from it. Here two names must be mainly cited (there are others of course): David Hilbert and his great disciple Hermann Weyl.

I may be allowed to illustrate this claim with a few very dry numbers. Let us look at Weyl's complete works, which fill four weighty volumes (quite apart from his books). In the first volume we find that no fewer than seventeen out of the twenty-nine papers treat subjects that later became directly or indirectly connected with Quantum Mechanics, especially wave mechanics. Indeed, twelve of the first thirteen papers, beginning with his "Inaugural Dissertation," deal with integral equations, differential equations, eigenfunctions, orthogonal function systems, questions of convergence, the Fourier-theorem, etc.

This series is interrupted only once by a short paper of four pages that deals with the definitions of fundamental mathematical concepts. The last paper of the series

---

[1] Originally published in *Exact sciences and their philosophical Foundations*. Proceedings of the Hermann Weyl Congress, Kiel 1985, Wolfgang Deppert, ed. (Frankfurt-Bern-New York-Paris: Peter Lang, 1985) pp. 161–190.

[2] *In seinem Artikel über Quantentheorie in der 1. Auflage des Blauen Handbuches (1925) vollbrachte er das Kunstwerk, unmittelbar vor der Auffindung der Quantenmechanik, also mit scheinbar noch unzureichenden Hilfsmitteln, praktisch jede physikalische Frage richtig zu beantworten* (Carl Friedrich von Weizsäcker, "Erinnerungen an Wolfgang Pauli," *Zeitschrift für Naturforschung* 14a (1959): 439–440).

studies the asymptotic distribution of eigenvalues (1911);[3] this problem is taken up again in a second series of five papers devoted to applications to physical problems. This series culminates in the paper "Über das Spektrum der Hohlraumstrahlung."[4] If we look now at the size of the papers, then this concentration on problems that we find treated later in Weyl's book is even more impressive; the papers mentioned make up about three-quarters of his first volume. And this is not all. We must remember that during this period Weyl wrote his celebrated first book, *Die Idee der Riemannschen Fläche.*[5] In this book at least two topics are treated that deal with what he will later use in the *Gruppentheorie*: the careful group theoretical discussions of the manifolds and the construction of H. A. Schwarz's universal covering surface, a tool that will bear great fruit later in his group theoretical investigations.

Hermann Weyl, 1885–1955

To the same period, and emerging of course from the preoccupation with these questions, belongs his obituary of Poincaré.[6] I mention it because what strikes us immediately is the enormous self-confidence that speaks from these lines, written by a young man twenty-six years old about the most famous mathematician of two generations before. Weyl had known of him through Poincaré's ambiguous relation with his teacher, Felix Klein, to whom he had dedicated his first book. Weyl, at the age of 26, speaks of Poincaré, albeit without any arrogance, as one speaks of an equal.

In the last article of this series, written in 1915, Weyl investigates the asymptotic distribution of eigenfrequencies of an elastic body.[7] The electromagnetic radiation field had been the simplest case; now he was to apply what he had learned to the next more general case that physics presented to him. To do this, Weyl had to acquaint himself with the theory of elasticity, a domain with which, then as today, not many professional physicists were familiar. The personal gain for him, besides becoming more deeply acquainted with physics, seems to have been a thorough grasp of tensor analysis, which proved useful later. The reader also notices that the exposition gains

[3] Weyl, *Gesammelte Abhandlungen,* vol. I, no. 13, "Über die asymptotische Verteilung der Eigenwerte," pp. 368–375.
[4] Weyl, *Gesammelte Abhandlungen,* vol. I, no. 18, pp. 442–460.
[5] Weyl, *Die Idee der Riemannschen Fläche* (Leipzig: Teubner, 1913). Engl. trans. *The Concept of a Riemann Surface,* Gerald R. MacLane, trans. (Reading, MA: Addison-Wesley).
[6] Weyl, *Gesammelte Abhandlungen,* vol. I, no. 15, pp. 390–392.
[7] Weyl, *Gesammelte Abhandlungen,* vol. I, no. 22, "Das asymptotische Verteilungsgesetz der Eigenschwingungen eines beliebig gestalteten elastischen Körpers," pp. 511–562.

in clarity, becomes more systematic, and that the crucial points of the proofs are stressed and exposed. Prof. Sternberg of Caltech wrote to me about this paper, "... Weyl's paper is often cited in the more mathematical literature on vibrations of elastic bodies and is generally viewed as a key contribution to this subject."

It may be mentioned that we find in this paper of 1915, for the first time, the term *Quantenhypothese*. Of course, we should not infer that only then did he learn of this new development. He had surely heard it in Göttingen long before, but his research then had been concentrated on purely mathematical questions; now we observe that his activities broaden more and more.

After this paper there is a gap of over ten years. In part, this is due to his military service during the war: indeed, the next paper on topological function theory and its relation to Hilbert's theory of class fields is written during that period. But the main reason for the gap is that General Relativity, created a few years earlier, now kept him busy and absorbed him for quite a while.

There is only one paper that calls forth our attention, that entitled "Das Verhältnis der kausalen zur statistischen Betrachtungsweise in der Physik," written in 1920.[8] There we find the following statements and questions:

> In what, besides causality, do the necessity and entitlement of statistics originate? Does it merely represent an abridged way to arrive at certain consequences of the laws of causality, or does it announce that no strict causal connection prevails in the world, but that "chance" beside law, is to be acknowledged as an independent power that restricts the validity of the law? The physicists are of the first opinion today.[9]

Written in 1920 (just after he had immersed himself in General Relativity) this is a remarkable statement and poses many questions. Weyl was not the first who saw the possibility of a science based in a fundamental way on the notion of probability (as the Copenhagen school indeed said a few years later). Franz Exner had apparently expressed this idea.[10] Whether Weyl was influenced by Exner or said it independently, as I suspect, I do not know. Did Weyl influence the scientists and especially the physicists? This may well be so, but I must leave the answer to specialists. For the record I may add a remark made by Otto Stern during a luncheon conversation at the home of Weyl's second wife.[11] Asked about the importance that one should attribute to the discovery of the probabilistic interpretation of Quantum Mechanics he cautioned, "in those years (1924 to 1927), everybody talked about probability."

---

[8] Weyl, *Gesammelte Abhandlungen,* vol. II, no. 38, pp. 113–122.
[9] *Woraus entspringt die Notwendigkeit und Berechtigung der Statistik neben der Kausalität? Stellt sie lediglich einen abgekürzten Weg dar, zu gewissen Konsequenzen der Kausalgesetze zu gelangen, oder zeigt sie an, daß in der Welt kein strenger Kausalzusammenhang herrscht, sondern der "Zufall" neben dem Gesetz als eine selbständige, die Gültigkeit des Gesetzes einschränkende Macht anzuerkennen ist? Die Physiker sind heute der ersten Ansicht* (Weyl, *Gesammelte Abhandlungen,* vol. II, p. 117).
[10] Exner is mentioned several times in Erwin Schrödinger's *Science and the Human Temperament* (1st ed., London: George Allen and Unwin, 1935; rpt. *Science, Theory and Man,* New York: Dover, 1957); see especially pp. xiv, xvii, 71, and 143.
[11] Editor's note. Weyl's second wife was the sculptor Ellen Lohnstein Bär, who he married in 1950, and who was the mother of David Speiser's wife Ruth Bär.

Then, of course, we find the four great papers on simple and semi-simple Lie-groups,[12] which Weyl later called his greatest achievement. With these papers we have now arrived at the years of the discovery of Quantum Mechanics.

The years 1925–26 are filled with a series of shorter papers on the orthogonal and the "complex" group (as he still called it at that time) and their representations. There is, of course, the celebrated paper written together with Fritz Peter.[13] Here he seems to have opened (I must follow his scientific biographers, Claude Chevalley and André Weil,[14] on this assertion) the new field of infinite dimensional representations. However, although all these topics have a direct or indirect contact with Quantum Mechanics, the new discoveries are not mentioned. It is true that we frequently find the name of Bohr, but it always refers to Harald and not to Niels!

But in 1927 he published a paper of 45 pages, "Quantenmechanik und Gruppentheorie."[15] Note the title: it contains the same words as his book, but *there* he will reverse the order and say *Gruppentheorie und Quantenmechanik.* This remarkable paper of 1927 is crucial to our understanding of how Weyl approached the new Quantum Mechanics. The paper is *not* directed towards the usual application of group theory to Quantum Mechanics (as, for example, to their spectra).

This application, as Weyl says in a footnote, had been initiated by Eugene Wigner.

Rather, as he says, his two-fold goal is:

1. to answer the question: how do I find the Hermitian form that represents a given quantity?
2. if I do have this form, what is its physical significance, and what can I learn from it?

The second question, he calls the easier of the two; and he can now, in virtuoso style, use for his purpose whatever mathematics he had developed in the earlier work.

The more difficult question, he says, is the first one.[16] John von Neumann and Pascual Jordan had made attempts in this direction, but neither satisfies him. His own attempt is based on the notion of a *pure state,* which, as he adds, von Neumann had also reached independently. The pure state is represented by a *ray* rather than by a vector. Thus, he now concentrates on ray representations of Abelian groups. Thus – and this must be stressed – Weyl found his *own,* independent access to the fundamentals of Quantum Mechanics. His result is contained in the following quotation:

> The kinematical structure of a physical system is expressed by an irreducible Abelian group of unitary ray rotations in system space. The real elements of the algebra of this group are the physical quantities of the system; the representation of the abstract group by rotations of

---

[12] Weyl, *Gesammelte Abhandlungen,* vol. II, no. 68, "Theorie der Darstellung kontinuierlicher halbeinfacher Gruppen durch lineare Transformationen I, II, III und Nachtrag," pp. 543–653.

[13] Weyl, *Gesammelte Abhandlungen,* vol. III, no. 73, "Die Vollständigkeit der primitiven Darstellungen einer geschlossenen kontinuierlichen Gruppe" (Fritz Peter and Hermann Weyl), pp. 58–75.

[14] Weyl, *Gesammelte Abhandlungen,* vol. IV, "Hermann Weyl (1885–1955) par C. Chevalley et A. Weil," pp. 655–686.

[15] Weyl, *Gesammelte Abhandlungen,* vol. III, no. 75, pp. 90–135.

[16] Prof. Mackey drew my attention to a slip in the manuscript.

system space associates with each such quantity a definite Hermitian form which "represents" it.[17]

We shall soon hear what he was to make of that. This investigation will later become the content of the last sections of the fourth part of the book and there it will be called "quantum kinematics."

I may mention, by the way, that we also find in this paper the perturbation expansion in terms of integral operators,[18] which was to play such a big role in renormalisation theory twenty years later.

The paper also shows Weyl in relation to the three fellow scientists who worked on the same or similar problems: Eugene Wigner, John von Neumann and Pascual Jordan.

Taking into account all these papers, we feel that he might well have chosen as the motto of his book a sentence that he wrote in the first of the four great papers: *Hier sollen alle diese Fäden zu einem einheitlichen Ganzen verwoben werden* (Here all these threads shall be woven into a uniform whole).[19]

***

Before coming to the book itself, I would like to take a short look at his later work. It is by no means only what Gottfried Keller (whom Weyl quotes in his book on classical groups) called *eine kleine Nachernte*, a small after-harvest. These papers stretch into his very last years, especially the one on eigenvalue problems which he applies again to the physics of radiation.

Concerning Weyl's later work in the theory of elasticity, Clifford A. Truesdell, who I had asked about the effect of Weyl's paper, wrote to me as follows:

> The paper on shock waves was of great importance. Weyl and Bethe, working unbeknownst to each other on classified projects during the war, succeeded in formulating the problem and going some way with it. The thermodynamic inequalities they proposed have been influential in many subsequent studies of finite waves ... There is an enormous subsequent literature. An idea of the continuing impact of the work of Bethe and Weyl can be got from the article by Peter Chen, in the second edition of my *Rational Thermodynamics* [Appendix 4A].

But then, we also find many papers on groups and algebras, for instance the famous one with Richard Brauer on "Spinors in *n* Dimensions,"[20] a much later one on the Dirac equation in General Relativity, and several on symmetry problems in

---

[17] *Der kinematische Charakter eines physikalischen Systems findet seinen Ausdruck in einer irreduziblen Abelschen Drehungsgruppe, deren Substrat der Strahlenkörper der "reinen Fälle" ist. Die reellen Größen dieses Gruppengebietes sind die physikalischen Größen; die Hermiteschen Matrizen, als welche sie vermöge der Darstellung der abstrakten Gruppe durch Drehungen erscheinen, sind die Repräsentanten der physikalischen Größen,* ... (Weyl, *Gesammelte Abhandlungen*, vol. III, p. 118; translation from *The Theory of Groups and Quantum Mechanics*, H. P. Robertson, trans., London: Metheun & Co., 1931; rpt. New York: Dover, 1950, p. 275).

[18] See Weyl, *Gesammelte Abhandlungen*, vol. III, p. 112.

[19] Weyl, *Gesammelte Abhandlungen*, vol. II, no. 68, "Theorie der Darstellung Kontinuierlicher halbeinfacher Gruppen durch lineare Transformationen," p. 547.

[20] Originally published in *American Journal of Mathematics* **57**, 2 (1935): 425–449; rpt. in *Gesammelte Abhandlungen*, vol. III, no. 105, pp. 493–516.

Quantum Mechanics.[21] The most surprising ones are those in which he applies (in the third one written together with Georg Rumer and Edward Teller) the theory of characters to the theory of chemical bonds.[22] Then there is the paper on symmetry,[23] later to be expanded into the Vanuxem lectures, to which I shall come back later.

Almost exactly in the middle of this great stream stands *Gruppentheorie und Quantenmechanik*, to which I will now turn.

***

The reader may ask what the lecture was on which the book is based, as he states in the preface, and in which circumstances it was delivered. In the preface to the first edition Weyl writes,

> In the winter term of 1927/28, Zurich suddenly found itself, by the simultaneous departures of Debye and Schrödinger, deprived of all theoretical physics. I tried to fill the breach by transforming an announced course on group theory into one on group theory and quantum mechanics. The manuscript of this course, worked out by F. Bohnenblust, is the basis of the book that the reader holds in his hands."[24]

The book is dedicated to his friend, Walter Dällenbach.

Through the courtesy of Drs F. Schindler and B. Glaus of the Eidgenössische Technische Hochschule, or ETH, the Swiss Polytechnical Institute, I am able to add a few details. The title of the course was originally: "Kontinuierliche Gruppen und ihre Invarianten." If you look at the book, you find about sixty sections, each written in a compressed style. So, you expect a course of at least sixty, if not a hundred and twenty lectures. Remember that a lecture in Switzerland, as in Germany, lasts only 45 minutes.

But, in fact, it was a *two lecture per week course of one semester only,* that is, approximately thirty lectures! So Weyl can have talked about only a small part of what he wrote. Which part it was, and how he made the selection, I do not know.

How big was the audience? Dr. Schindler found out that the total number of students for all four years in the department of mathematics and physics was twenty-nine! As the academic year at the ETH starts in the fall, only students in the last two years were able to follow him.

---

[21] Weyl, *Gesammelte Abhandlungen,* vol. III, no. 83, "Gravitation and the electron," pp. 217–228; no. 84, "Gravitation and the electron," pp. 229–244; no. 85, "Elektron und Gravitation," pp. 245–267; no. 86, "The spherical symmetry of atoms," pp. 268–281; vol. IV, no. 141, "A remark on the coupling of gravitation and electron," pp. 286–288.

[22] Weyl, *Gesammelte Abhandlungen,* vol. III, no. 90, "Zur quantentheoretischen Berechnung molekularer Bindungsenergien," pp. 308–317; no. 91, "Zur quantentheoretischen Berechnung molekularer Bindungsenergien II," pp. 318–324; no. 97, "Eine für die Valenztheorie geeignete Basis der binären Vektorinvarianten" (Georg Rumer, Edward Teller und Hermann Weyl), pp. 380–385.

[23] Weyl, *Gesammelte Abhandlungen,* vol. III, no. 111, "Symmetry," pp. 592–610.

[24] *Im Wintersemester 1927/28 sah sich Zürich plötzlich durch die gleichzeitige Wegberufung von Debye und Schrödinger jeglicher theoretischen Physik beraubt. Ich versuchte in der Bresche zu springen, indem ich eine angekündigte Vorlesung über Gruppentheorie in einer solche über Gruppentheorie und Quantenmechanik umwandelte. Die von Herrn F. Bohnenblust angefertigte Ausarbeitung dieser Vorlesung bildet den Grundstock des Buches, das der Leser in Händen hält* (Hermann Weyl, *Gruppentheorie und Quantenmechanik,* Leipzig: S. Hirzel, 1931, p. iii).

Hermann Weyl, ca. 1930
(Göttingen). Courtesy of the
Archives of the Mathematisches
Forschungsinstitut Oberwolfach

Of this half, take a third, add a few postgraduates and a handful of students from the university where Schrödinger had left. My guess is that there were about six, perhaps ten students, that is: at the beginning at least! I do not know anyone still alive who can tell us about these lectures.

Weyl also appears in the *Vorlesungsverzeichnis*, the roster of lecturers, in a solid Swiss setting. Alphabetical order placed him between Herr Professor Weber and Herr Professor Wiegner. But this Professor Weber is not Heinrich Weber, the physicist and teacher of Einstein. He is Herr Oberstkorpskommandant Weber, who lectures to career officers on *Feld- und permanente Befestigungen.* Professor Wiegner is not Eugene but G. He writes his name not only with an "i" but with "ie," and does not lecture on group theory but on *Die Chemie der Milchprodukte* and he teaches *Kapitel der Fütterungslehre (mit Übungen).*

Between the two stands Professor Hermann Weyl, Bolleystraße 52. Bolley, incidentally, was Pompeius Bolley, like Weyl a German who had become a professor at the ETH, but in chemistry; I am sure that, like almost everybody who has a street named after him in Zurich, he was a great pedagogue!

The second edition is based on lectures delivered at Princeton University in 1928–29. There Weyl enlarged the third and the fourth parts and entirely rewrote, as he tells us, the fifth part. He says, "the lectures which I gave there and in other American institutions afforded me a much desired opportunity to present anew, and from an improved pedagogical standpoint, the connection between groups and quanta."[25] H. P. Robertson, who had begun the translation of the first edition, completed it, based on the second.

***

Now to the book itself. It is organized in five chapters, dedicated respectively to the following five topics:

I. Linear unitary spaces (Hilbert spaces);
II. The physical foundations of Quantum Mechanics and its logical and mathematical structure;
III. Group theory;
IV. The application of group theory to Quantum Mechanics as a tool as well as a way to formulate the underlying logical structure;
V. The symmetric permutation group, the group algebra and its relation to the tensor and representation theory of the linear group.

---

[25] Weyl, *The Theory of Groups and Quantum Mechanics,* H. P. Robertson, trans. (London: Metheun & Co., 1931; rpt. New York: Dover, 1950), p. ix.

GRUPPENTHEORIE UND QUANTENMECHANIK

VON HERMANN WEYL

ORD. PROFESSOR

DER MATHEMATIK AN DER EIDGENÖSSISCHEN

TECHNISCHEN HOCHSCHULE

ZÜRICH

VERLAG VON S. HIRZEL IN LEIPZIG · 1928

Title page, first edition of Weyl's *Gruppentheorie und Quantenmechanik*. Courtesy of the Biblioteca "G. Peano", Department of Mathematics, University of Turin

The size of the five chapters increases in this order, as does their difficulty. The first is the shortest, the last one (even in the first edition) the longest; it is the most difficult by far but also the most original.

Chapter I

A few remarks concerning the first chapter must suffice. Recall that Weyl began his book *Raum, Zeit, Materie (Space Time Matter)*[26] with a presentation of geometry in vector form based on an axiomation in the sense of Hilbert that was due to himself. At the beginning of this book we find again (he expresses himself with regrets that he *must* respect it) a formal treatment of vector calculus; but this time the spaces are complex. What strikes us at once, if we read these approximately thirty pages, is that the approach and the treatment are still entirely valid today. Take, for instance, Paul Halmos's little book on vector space;[27] we find many more details here, and much is smoothed out, but the spirit and scope are the same. A treatment of affine geometry, including dual space, precedes the treatment of metric space and *only then* are operators introduced. Today, a modern treatment is usually more abstract, often simpler, and sometimes more transparent; but substantially little is added that is new.

In the last section he sketches the generalization to infinite dimensional spaces with which he was, as we saw, acquainted through his work on integral and differential equations. The decisive steps taken later, the extension to unbounded operators, seem due to von Neumann, with whom Weyl was in close contact at that time. But the fact that Weyl starts throughout from the finite case has this happy effect, that the ε-garbage is never allowed to obscure the algebraic structure, which remains central throughout. The only essential point that one may miss concerns square summable sequences and square summable spaces, i.e., the l^2 and L^2 spaces. This was done later by von Neumann.

Most physicists must have been baffled to find that, in a book where they had hoped to learn about atoms and molecules, they first had to learn the mathematics of linear spaces! But luckily, as I shall show, a few of them understood at once the deeper sense of Weyl's procedure.

Chapter II

It is the second part, Weyl's presentation of the fundamentals of Quantum Mechanics, that will above all attract the physicists' attention.

How does Weyl introduce his subject? He begins, naturally enough, with a short historic introduction retracing the development from Planck and Einstein to Bohr, the quantization rules and its experimental confirmations.

But the important experimental result is the *Ritz-Rydberg combination principle*, which he discusses at length and which he shows to be a direct consequence of Heisenberg's new mechanics. There follows the remarkable sentence:

> Another approach to quantum mechanics was discovered by *L. de Broglie* and *E. Schrödinger*. This approach seems to me less cogent, but it leads more quickly to the fundamental principles of quantum mechanics and to the most important consequences for experimental science. We shall therefore follow it, since we are more concerned with

---

[26] Hermann Weyl, *Raum, Zeit, Materie, Vorlesungen über Allgemeine Relativitätstheorie* (Berlin: Springer, 1919); *Space Time Matter,* Engl. trans. from the 4th ed. by Henry Brose (London: Methuen & Co., 1922; rpt New York: Dover, 1952).
[27] Paul R. Halmos, *Finite-Dimensional Vector Spaces* (New York: Springer, 1974).

giving a short but comprehensive account than in giving a complete discussion of the physical foundations.[28]

This statement is not only noteworthy as the opinion and the judgment of a contemporary scientist; it shows that it was not the fact that the mathematics, which Weyl understood and knew by heart, become at once applicable that convinced him of the validity of the new Quantum Mechanics, but rather it was the "cogency" of its conceptual basis together with the fact that the Ritz-Rydberg principle covered such an immense experimental material. Thus, the central sections of chapter II are 7, 8 and 11.

That Weyl did not choose the easy way – namely to use the lecture as a pretext for presenting his own mathematical results and to remain on the surface of Quantum Mechanics – is a testimony not only to his intellect but also to his strength of character. The fact that he was able to learn Quantum Mechanics in such a short time and to master the subject from all sides cannot be stressed enough. Consider how long – even today where we have comprehensive text books – it takes a student to go as far as Weyl did! It is a real triumph for mathematical thinking!

This is probably the reason for another accomplishment where I believe Weyl holds a unique position. It is fair to say that general relativity aroused the interest of mathematicians as much as or even more so than that of physicists. Not so with Quantum Mechanics! Mathematicians only exceptionally took to it, and those who did, I suspect, came to it through Weyl's book. The penetration and thoroughness with which the structure of Quantum Mechanics is analyzed, illustrated with examples taken from physics and pursued even into minute details of the formalism, is striking. I have stressed Weyl's parsimony with words when presenting formal matters. Only when he discusses fundamental and, for once we may be allowed to say, revolutionary ideas and questions, does his style become broader. Even then, I would not call him "loquacious," nor even "wordy;" but the reader is not denied *ein gründliches sich Umsehen*, a thorough look around. Besides Heisenberg's equations, it is now his own analysis of measurements which led him, as we have seen, to the "pure state," a notion that was generalized by von Neumann, to the mixture that provides the fundament.

Here, I feel, is the moment where the indebtedness of today's readers to *Weyl the historian,* must be mentioned and stressed. Weyl's book, as he notes, was almost the first systematic presentation of the new Quantum Mechanics. Thus, he had to work from original papers. He performed this part of his job very conscientiously, almost like an archivist, one could say. His list of references is large; it is a real treasure house, and to *Weyl the mathematician* and *Weyl the physicist* (about both of whom more will be said later), we must add *Weyl the historian of science.* The entries in the notes and references are extremely valuable. There too, we can see that Weyl made himself thoroughly acquainted with a great number of the publications of the period. It is amazing how much physics Weyl has absorbed in almost no time. How was this possible?

---

[28] *Einen anderen Zugang zur Quantentheorie entdeckten L. de Broglie und E. Schrödinger. Er erscheint mir weniger zwingend, führt aber rascher zu der neuen Grundauffassung der Quantenmechanik und zu den wichtigsten an der Erfahrung zu kontrollierenden Ergebnissen. Darum wollen wir ihm folgen, zumal es uns mehr um eine kurze zusammenfassende Darstellung als um eine lückenlose Begründung der physikalischen Grundlagen zu tun ist* (Hermann Weyl, *Gruppentheorie und Quantenmechaniks*, 2nd ed., p. 43; Engl. trans. *Theory of Groups and Quantum Mechanics*, p. 48).

Earlier I mentioned Weyl's own mathematical investigations, which led him to his understanding of Quantum Mechanics. He must have recognized *at once* both the mathematical content and the deeper significance of so many results, while most physicists had to take great pains and a long time to see them. Yet, the mathematical formalism, as such, is never allowed to dominate the presentation. If we read Weyl's book today, the impression that we receive is on the one hand, one of extreme economy, even austerity, with respect of the formalism, and on the other hand, one of great conciseness. As a test, take any early book on Quantum Mechanics, or for that matter even a new one, and see who heaps the least number of mathematical details concerning, say, Laguerre polynomials on the reader's head – the author of this book or Weyl? The answer will almost always be Weyl, who is usually more successful in suppressing merely formal knowledge of secondary importance. On the other hand, where it *is* necessary – for instance, in the notoriously tricky theory of collision which he treats following Max Born – he does not neglect anything essential.

Most physicists must have found this book difficult, even puzzling. Indeed, even today, the whole book, especially chapter II, is not a text book in the ordinary sense; it is not for beginners, but for the advanced scientist. But this is not due to a "lack of physics" etc., as is said sometimes, but rather to the economy of its presentation. Its motto might be *In der Beschränkung zeigt sich der Meister.*[29] And there is, of course, another, deeper reason.

The principal characteristic of Weyl's presentation of Quantum Mechanics seems to me to be the following: most books on Quantum Mechanics, at least those of the early period, mainly explain the properties of atoms, molecules, electrons, etc. and perhaps the radiation field as well. These properties, of course, are linked by a formalism, but the structure of the link itself usually receives little attention. Not so in Weyl's book! *Here the structure itself is also made the center of an investigation* and the result of this investigation becomes for us a source of deeper understanding. The applications then follow naturally, the simple ones immediately, while the more complicated ones will be handled with the group theoretical tools which he will now put at the physicists' disposal. In this, as we shall see, Weyl immediately found at least one great follower.

I would like to add a comment on a particular statement made by Weyl in the second part of the book on the Hamilton equations. He writes:

> It is a universal trait of Quantum physics to retain all the relations of classical physics; but whereas the latter interpreted these relations as conditions to which the values of physical quantities were subject, in all individual cases, the former interprets them as conditions on the quantities themselves, or rather on the Hermitian matrices which represent them.[30]

---

[29] *In der Beschränkung zeigt sich erst der Meister, / Und das Gesetz nur kann uns Freiheit geben* (None proves a master except by limitation, / And only law can give us liberty) Johann Wolfgang von Goethe, *Natur und Kunst*, Engl. trans. by Michael Hamburger, in *Goethe: The Collected Works, Vol. 1: Selected Poems* (Princeton: Princeton University Press, 1994), p. 165.

[30] *Es ist ein durchgehender Zug der Quantenphysik, daß sie alle Beziehungen der klassischen Physik aufrecht erhält; während diese aber sie deutete als Bindungen, denen die Werte der physikalischen Größen in allen möglichen Einzelfällen unterworfen waren, fordert jene ihre Gültigkeit für die Größen selber oder für ihre repräsentierenden hermiteschen Matrizen* (Hermann Weyl, *Gruppentheorie und Quantenmechaniks*, 2nd ed., p. 84; Engl. trans. *Theory of Groups and Quantum Mechanics*, p. 95).

Today, we look at Hamilton's mechanics differently; the work of Dirac, Jost, Souriau, Arnold and Gilon[31] taught us to look at Hamiltonian mechanics from the point of view of symplectic geometry, exactly as – following Weyl, von Neumann and Dirac – we look at Quantum Mechanics from the point of view of unitary geometry. We now see that Weyl did point out a serious defect, *not,* as he thought, of classical mechanics, but of our then prevailing understanding of it. Once more, Quantum Mechanics had overtaken Classical Point Mechanics, which took this step only much later.

Chapter III

Chapter III is again mathematical: it contains the introduction to group theory. As this was the original purpose of the lectures, it was perhaps the first part written down.

What I said about chapter II may be repeated here. Chapter III too is not really an introduction, certainly not one for beginners. Weyl goes (one is tempted to say "jumps") from one essential point to the next, leaving aside everything unessential and filling in details only sparsely. Sometimes, a section even remains sketchy. The reader is well-advised to have a pencil and a pad near by. On the other hand, Weyl is not above discussing examples, though he does not do so very often. The chapter begins with a confrontation of the axioms of group theory with the three equivalence relations. Following his teacher Felix Klein, I suppose, he takes transformation groups as his starting point and only then does he introduce abstract groups.

Besides Klein, Weyl seems have been guided by the work of Frobenius and that of Issai Schur (who is often mentioned and to whose memory Weyl later dedicated his fourth major book, *Classical Groups*), as well as by Emmy Noether for the group algebra.

Weyl treats discrete groups and Lie groups side by side, and shows that the theory of representations and characters is illustrated by examples of both branches. I point out especially the section on invariants and covariants that sketches ideas that he will pursue later in his *Classical Groups*. And, of course, the last section on "ray representations," to which I shall come back, must be mentioned.

The list of books referred to in this chapter begins with Andreas Speiser's *Theorie der Gruppen endlicher Ordnung,* the second edition of which had been published in 1927.[32] Speiser, who was also in Zurich then, had studied in Göttingen at the same time as Weyl. Together with Max Born they had edited Hermann Minkowski's complete works under the direction of David Hilbert.[33] Although they worked in Zürich on different fields, they remained in personal contact. When Schrödinger began to formulate de Broglie's ideas mathematically, he asked Speiser about integral equations. Speiser referred him to Weyl at once. "Oh no, I shall not go to Weyl" Schrödinger exclaimed, as Speiser told me: "he will guess at once what I am up to do." But later, apparently, he went to Weyl anyway!

---

[31] Paul A. M. Dirac, *The Principles of Quantum Mechanics,* 4th ed. (Oxford: Clarendon, 1958); Res Jost, "Poisson Brackets," *Reviews of Modern Physics* **36**, 2 (1964): 572–579; J. M. Souriau, *Structure des systèmes dynamiques* (Paris: Dunod, 1970); Vladimir I. Arnold, *Mathematical Methods of Classical Mechanics* (1978), 2nd ed., (Heidelberg: Springer, 1989). Philippe Gilon, "Etude des variétés emplectiques et géométrisation du flot hamiltonien", Ph.D. thesis, Université Catholique de Louvain, 1984.

[32] Andreas Speiser, *Theorie der Gruppen von endlicher Ordnung* (first ed. Berlin: Springer, 1923).

[33] *Gesammelte Abhandlungen von Hermann Minkowski* 2 vols., A. Speiser and H. Weyl, eds., Introduction by D. Hilbert (Leipzig, 1911; rpt. New York: Chelsea Publishing, 1967).

Speiser had also instigated two of his students, J. J. Burckhardt and W. Schubarth, to translate Dickson's *Algebras*,[34] also mentioned in Weyl's list and certainly of great importance for the book, for, as we shall see, group algebra plays a central role in its last part. When Weyl's book appeared, Speiser's had already been used by Hans Bethe for computations on quantum mechanical problems of lattices.[35]

Chapter IV

The fourth chapter presents the application of group theory to Quantum Mechanics: it is without a doubt the chapter which attracted the greatest attention of the physicists. It is divided into four sub-chapters:

A. The Rotation Group;

B. The Lorentz Group;

C. The Permutation Group;

D. Quantum Kincmatics.

Not much need to be said about the first section, but a few points that strike the reader may be mentioned.

From the start, Weyl's treatment of the rotation group is in matrix formulation and based on SU(2) rather than on SO(3). This not only makes a unified treatment possible, but it also has the advantage that when Weyl explains, for instance, the anomalous Zeeman effect, the accent is *not* placed, as is so often the case, on the new mathematics of the spinors. These were usually presented in those days and even much later as an exotic curiosity, if not a monstrosity. Instead, with Weyl, the reader has learned it from the start, the attention is now on the physics and together with Weyl's thorough treatment of all physical questions, it deserves our unbounded admiration. Not only for the way in which such a thorny subject as Landé's rules is derived and displayed, but also for the fact that the restrictions under which a derivation is made, and the inequalities that must hold, but are so often swept under the carpet, are always explicitly mentioned here. Besides, there are many hidden treasures; the Wigner-Eckhardt theorem is implicitly present and, of course, through his paper on Lie groups, Weyl is aware of the very special character of SU(2). The possibility of "multiple weights" is (implicitly) stated. I mention this, because even more than thirty years later, many physicists (like myself, but even those more learned) were puzzled by them when they came across them for the first time.

The sections on the Lorentz group and on the Dirac equation are among those that strike the reader most, for Weyl does *not* use the Dirac matrices! He had discovered, following Dirac, his own "Weyl equation" with the two-component spinors that is invariant under the proper Lorentz group. So, Weyl formulated everything in terms of the two-component spinors, in a way that was altogether unusual for a long time. It would be only thirty years later, when the violation of parity[36] and the exact laws of weak interactions[37] were discovered, that the importance of the two-component

---

[34] Leonard E. Dickson, *Algebras and their Arithmetics* (Chicago: University of Chicago Press, 1927); German trans. *Algebren und ihre Zahlentheorie. By L. E. Dickson. Mit einetn Kapitel über Idealtheorie von A. Speiser* (Zurich und Leipzig, Orell Füssli Verlag, 1927).

[35] H. Bethe, private communication.

[36] T. D. Lee and C. N. Yang, "Question of Parity Conservation in Weak Interactions," *Physical Review* **104**, 1 (1956): 254–258.

[37] See R. E. Marshak, Riazuddin, C.P. Ryan, *Theory of Weak Interactions in Particle Physics* (New York: Wiley Interscience, 1969); W. R. Theis, "Eine Rangordnung der einzelnen Terme in den schwachen Wechselwirkungen," *Zeitschrift für Physik A Hadrons and Nuclei* **150**, 5

spinors was understood. Then we find a careful investigation of space inflection and time reversal. The implied existence of the antielectron is stated but rejected by him on grounds of the *experimental* situation. Surely, if physicists like to insist on the "formal" character of the book, which sometimes misses "physics," they must acquit a mathematician if he puts too much faith in their pronouncements and they must acknowledge their own responsibility!

Here is the place to note another gap in Weyl's book, a gap that we find wherever we look into the writings of those years. Weyl, like everybody else, does not mention, let alone treat, the *Galilei group.* At that time the idea of relativity was still fixed to Einstein's discovery and therefore to the Lorentz and Poincaré groups. This misapprehension then, of course, contributed to the lack of understanding of the true significance of the Dirac equation. We find a similar lack of clarity even in the otherwise extremely penetrating and very illuminating correspondence between Ehrenfest and Pauli.[38] There Ehrenfest complains that Schrödinger's theory is not really a *Nahewirkungstheorie*, a theory where all effects propagate locally through space and time. But, of course, forces that act at a distance *are* the natural ingredient of the Schrödinger equation, since it is Galileo invariant. But this point would only be cleared up much later, I think first by V. Bargmann.[39]

The third section contains the application of permutation theory to the spectra, together with one of the most lucid explanations of Pauli's exclusion principle. Weyl's abstract formulation makes this chapter difficult to read, but it assures the full generality of his formulations.

I have already mentioned Weyl's paper in which he published his discoveries on the pure state and on quantum kinematics as well as his exchange with John von Neumann. Here they are presented again, in great detail.

J. von Neumann later formulated, together with Garret Birkhoff, a rigorous axiomatic foundation of Quantum Mechanics, where he showed the radical difference between Classical and Quantum Mechanics. Von Neumann and Birkhoff made it explicitly clear that Quantum Mechanics is not just a new physical theory with a new and unusual dynamics, but rather, Quantum Mechanics is conceived *in a new, non-Boolean, logic,* fundamentally different from the logic of classical physics, one that can be best understood if Quantum Mechanics is understood from the notion of the "pure state," just as Weyl had done.

Their paper did not attract much attention when it was written, but it did catch *Weyl's* attention! Weyl later mentioned it in his article "The Ghost of Modality."[40] Only around 1960, after Weyl's death, was interest in it revived by David Finkelstein, and from then on, quite a number of people worked in this direction. But, even then it did not penetrate deeply into the physics community. This is regrettable,

---

(1958): 590–592; R. P. Feynman and M. Gell-Mann, "Theory of the Fermi Interaction," *Physical Review* **109** (1958): 193–198; E. C. G. Sudarshan and R. E. Marshak, "Chirality Invariance and the Universal Fermi Interaction," Proceedings of the Padua–Venice Conference on Mesons and Recently Discovered Particles (1957) and *Physical Review* **109** (1958): 1860–1862.

[38] See Wolfgang Pauli, *Wissenschaftlicher Briefwechsel mit Bohr, Einstein, Heisenberg u.a.*, vol. 2. Sources in the History of Mathematics and Physical Sciences (Berlin, Heidelberg: Springer: 1985).

[39] V. Bargmann, "On unitary ray representations of continuous groups," *Annals of Mathematics* **59**, 1 (1954): 1–46.

[40] Hermann Weyl, "The Ghost of Modality," *Philosophical Essays in Memory of Edmund Husserl* (Cambridge, MA: Harvard University Press, 1940). Rpt. *Gesammelte Abhandlungen*, vol. III, no. 118, pp. 684–709.

I think; had this been the case I am convinced that many discussions on the interpretation of Quantum Mechanics, etc. would have been simplified and even recognized as superfluous. Furthermore, I am sure that our understanding of Quantum Mechanics would have been deepened and the teaching of it would have gained in clarity.

Chapter V

Finally, a few words on the fifth chapter of the book, its most original part.

I remember that when I read the book for the first time as a student, I had to capitulate here, and not only partially: of this part I could grasp only very little. This, however, was not due only to my ignorance and my lack of intelligence, extensive as they both were. Weyl himself had already recognized his share in my, and presumably most people's, failure, when in 1929, he rewrote the fifth part entirely. But, alas, this second edition had not been bought in Basel and I did not know of its existence! Later, when I worked on internal symmetries, I had many reasons to regret not having mastered the fifth part of the book. There is no doubt that in its new form the presentation gained immensely in clarity. In 1929, Weyl had published a paper, "Der Zusammenhang der linearen und symmetrischen Gruppen,"[41] apparently following an idea of Schur. Obviously, here he had found the key that had thus far been missing. Out of group algebra, everything seemed to flow straight away and to fall into place.

He shows that it is the relationship between the group of linear transformations and the permutation group, discovered by Frobenius, that yields the key to our understanding of the atoms and molecules. It is here that he introduces his "musical notation" – the "sharp" # and the "natural" ♮ – that lead back and forth from one algebra to the other. It has been said that modern mathematics tries to replace computing as much as possible by thinking and reflection; this, at any rate, is what Weyl does here for physics and for chemistry.

This fifth part is the climax of the book; *here* the mathematics are pushed the furthest and *here* we find the most advanced and the most refined applications to physics *and* indeed to *chemistry*! Moreover, as the table of contents already shows, the two are more intertwined there than anywhere before in the book. It is organized in three sub-chapters:

A. the general theory;
B. the physical applications;
C. the explicit algebraic construction.

With respect to the applications, I may point to what comes as the biggest surprise – his thorough exposition of the theory of valences and of chemical bonds, linked to the theory of Young patterns and characters. I may mention here something that I have learned from a paper of Weyl himself[42] and from his obituary of Emmy Noether.[43]

In the 1880s, two mathematicians, J. J. Sylvester and P. A. Gordan, had sensed the connection between the theory of chemical valences and the theory of binary invariants. Gordan, apparently an extremely original mind as well as character, thereupon demanded that all German universities establish chairs for theoretical chemistry at once! Weyl comments that several decades later these speculations

---

[41] Weyl, *Gesammelte Abhandlungen,* vol. III, no. 79, pp. 170–188.
[42] Weyl, *Gesammelte Abhandlungen,* vol. III, no. 91, "Zur quantentheoretischen Berechnung molekularer Bindungsenergien II," pp. 318–324.
[43] Weyl, *Gesammelte Abhandlungen,* vol. III, no. 102, pp. 425–444.

eventually found full jusitification! I am ignorant as to whether historians have followed up Sylvester's and Gordan's work and its possible influences, but, if not, then I hope that it will be investigated with all possible energy and care!

Indeed, Weyl was not only the "messenger," as he said, between mathematicians and physicists but also between mathematicians and chemists.

This short survey cannot do justice to this chapter, which, in my opinion, still provides stimulation for new research today.

***

What is new in the book? I have tried to give some indications of this with respect to physics and chemistry. With respect to what is new in mathematics, I do not feel competent enough to speak. A long and penetrating investigation, like the one that we owe to Chevalley and Weil,[44] by someone equally familiar with the entire body of literature of the 1920s, would be needed. I hope that we shall have it one day. Its result will certainly fill more than one talk. The two talks given at the 1985 Weyl conference by Profs. Hans. Freudenthal and Gerhard Mack go a long way towards filling this gap.[45]

An evaluation of what is original is also made difficult by the fact that Weyl attacks everything from his own point of view, and translates it into his own language. If one catches his idea, one may swim comfortably along; if not, one is better off to take only his hints and to try to redevelop the results oneself by working out a simple example. What stands often enough between him and the physicists are the abstract formulations. But, of course, this is precisely what makes his proceeding so powerful!

Reading Weyl now, I am reminded again of how my teacher, Markus Fierz, described him to me: "the man who makes all simple things complicated and all complicated things simple." In the end, the second positive part, of course, remains decisive. "Weyl always takes up the problem at its root," Andreas Speiser once remarked.

The reader who works out examples himself will be rewarded by making his own discoveries. Often, he will then underestimate Weyl's stimulus and overestimate his own contributions. But, so what? He will soon return to a more sober evaluation.

***

How was the book received by scientists?

With respect to its reception by physicists, I may first quote from the Pauli correspondence. Wolfgang Pauli, it may be recalled, had joined Weyl in Zurich at the ETH shortly after the discovery of Quantum Mechanics. Pauli was certain that Weyl had had a hand in this appointment: "I think he liked my spinors," he told me once, referring to his discovery of the Pauli equation. Pauli had written Weyl a long letter[46] on his paper, "Quantenmechanik und Gruppentheorie," where, although he takes him to task for a mathematical mistake, he expresses his interest and his agreement. "The paper interested me very much and I even believe that I understood its essentials" he

---

[44] *Op. cit.*, in Weyl, *Gesammelte Abhandlungen,* vol. IV, pp. 655–686.

[45] See Gerhard Mack, "Problems of Quantized Gauge Fields," and Hans Freudenthal, "Hermann Weyls Beitrag zur Theorie der Darstellung der Lie-Gruppen" in *Exact sciences and their philosophical Foundations.* Proceedings of the Hermann Weyl Congress, Kiel 1985, Wolfgang Deppert, ed. (Frankfurt-Bern-New York-Paris: Peter Lang, 1985).

[46] Pauli to Weyl, 29 January 1928 in Wolfgang Pauli, *Wissenschaftlicher Briefwechsel mit Bohr, Einstein, Heisenberg u.a.* Sources in the History of Mathematics and Physical Sciences 2 (Berlin, Heidelberg: Springer: 1985), letter no. 181, vol. I, p. 427, note 1.

said, with the touch of the Mephistophelian irony which he reserved for the famous ones among his correspondents.

Speaking about what we today call "Weyl's form of the uncertainty relations," he said: "I was very glad about your remark that the transitions from the finite to the infinite case leads first to the equations [in integrated form] and that [Heisenberg's form] (like the corresponding operator equations) is a secondary consequence."[47]

Pauli then explained the possible importance of this fact for the quantization of fields, with which he was busy at the time; and expressed his hope that the clarification he received from Weyl's paper would lead him further.

In a letter to Bohr, Pauli said, "From Weyl I have learned much group theory, such that I can now really understand Wigner's papers as well as those of Heitler and London, which are important also for Heisenberg's new theory of Ferromagnetism."[48] Thus, unlike many others, Pauli understood from the start the reach of Weyl's endeavour.

For the second edition, the editor added a series of quotations from different reviews of the first printing. Two may be quoted. The first one is from the Belgian physicist Charles Manneback, who knew Weyl and read his book, it would seem, with particular care. He wrote in the *Revue des Questions Scientifiques*:

> *C'est la seconde fois que l'auteur, dont le nom s'est imposé depuis longtemps dans le domaine des Mathématiques pures et de la Philosophie scientifique, publie un ouvrage de Physique théorique.*
>
> *L'ouvrage de M. Weyl rend inutile la lecture de nombreux mémoires originaux tant physiques que mathématiques; il donne un fil conducteur au milieu de la production théorique. Il est indispensable à toute personne soucieuse de suivre la physique nouvelle dans des exposés de première main. Comme l'auteur le dit, les chapitres physiques ont été rédigés surtout pour lecteurs mathématiciens, les chapitres mathématiques pour les physiciens. Cet ouvrage se caractérise par son unité, sa forte charpente et la généralité des points de vue.*

Three years later, Manneback wrote about the second edition: *Comme on le sait déjà, le traité du professeur Weyl est un ouvrage fondamental sur la théorie des quanta.*

Surprising is the echo in the *Zeitschrift für angewandte Chemie*. The reviewer, K. Bennewitz, writes:

> *Was hier geboten wird, ist mehr als eine bloße Zusammenschrift eines geistigen Forschungsgebietes, das im Laufe dreier Jahre einen unerhörten Aufschwung genommen hatt; es ist eine fein durchdachte Neuschöpfung geworden. Wem es Ernst ist um das Verständnis unserer geistigen Entwicklung, dem wird das Werk viel Genuß bereiten, vorausgesetzt, daß er den Sinn für abstrakt-mathematische Lehren besitzt.*
>
> *So schwer zugänglich dem Fernerstehenden diese Dinge sein mögen, so scheint eine zukünftige Physik ohne sie nicht mehr möglich; dem*

---

[47] The words in the [ ] brackets are by the author; they replace the formulae in the letter.
[48] Pauli to Bohr, 14 July 1928 in Wolfgang Pauli, *Wissenschaftlicher Briefwechsel, op. cit.*, letter no. 203, vol. I, p. 465.

> *Chemiker aber öffnen sie die Augen über das Wesen seiner Materie, über das periodische System, die Molekülbildung und die Valenz.*[49]

H. B. G. Casimir writes in his book *Haphazard Reality*:

> But, for me, the great revelation was Hermann Weyl's *Gruppentheorie und Quantenmechanik*. I think it is a wonderful book. I had – and still have – the feeling that this is the right way to formulate mathematics, that the concepts Weyl formulates have just the right level of abstraction, yet correspond to something "real," whatever that may mean in pure mathematics. If I have later been able to apply group theory with some measure of success to several problems, and if I have even been able to make a small but not entirely insignificant contribution to that field of pure mathematics, this is thanks to Weyl's book. I never met Weyl until after the Second World War at Princeton, when, like Einstein, he had found a position at the Institute of Advanced Studies. Then I was happy to be able to tell him how grateful I was. (Weyl on that occasion paid me a probably wholly undeserved compliment: he introduced me to a colleague as a physicist with the soul of a mathematician.)[50]

In a letter to me, Casimir wrote: "my thesis Rotation of a Rigid Body in Quantum Mechanics (it contains the "Casimir-operator") is partly inspired by Weyl."

But probably the most important tribute, with respect to the future development of physics, is that of Dirac, who wrote in the first edition of his celebrated book:

> With regard to the mathematical form in which the theory can be presented, an author must decide at the outset between two methods. There is the symbolic method, which deals directly in an abstract way with the quantities of fundamental importance (the invariants etc., of the transformations) and there is the method of coordinates or representations, which deals with sets of numbers corresponding to these quantities. The second of these has usually been used for the presentation of quantum mechanics (in fact it has been used practically exclusively with the exception of Weyl's book *Gruppentheorie und Quantenmechanik*). It is known under one or other of the two names 'Wave Mechanics' and 'Matrix Mechanics' according to which physical things receive emphasis in the treatment, the states of a system or its dynamical variables. It has the advantage that the kind of mathematics required is more familiar to the average student, and also it is the historical method.
>
> The symbolic method, however, seems to go more deeply into the nature of things. It enables one to express the physical laws in a neat and concise way, and will probably be increasingly used in the future as it becomes better understood and its own special mathematics gets developed. For this reason I have chosen the symbolic method, introducing the representatives later merely as an aid to practical calculation. This has necessitated a complete break from the historical

---

[49] *Zeitschrift für angewandte Chemie* **42**, 19 (11 Mai 1929): 486.
[50] H.B.G. Casimir, *Haphazard Reality. Half a Century of Science* (New York & London: Harper & Row, 1983), p. 76.

line of development, but this break is an advantage through enabling the approach to the new ideas to be made as direct as possible.[51]

Weyl's name is, at least in the part retained in the later editions, the only one mentioned in the foreword. Thus did Dirac attest to the importance of Weyl's book for his own supreme achievement, where, even more so than in Weyl's, ordinary Quantum mechanics and Relativistic Field theory appear from the root as one and the same science.

It is interesting to compare Weyl's access to Quantum Mechanics with Dirac's. Both were captivated by Heisenberg's discovery. But then, it was the *mathematician* Weyl, who, as we saw, became convinced by the straightforward way the *physical* Ritz-Rydberg principle could be formulated and explained in this frame; and it was the *physicist* Dirac who noticed the *mathematical* analogy between Heisenberg's commutation relations and Poisson-brackets! Of course, throughout Dirac uses vectors, not rays, for his formulation.

Here I may add a regret: namely, that Weyl never returned to his masterpiece to give us *eine umgearbeitete dritte Auflage*, a revised third edition, which would have shown, even more than the first two, Quantum Mechanics as a whole, beginning anew with "Quantum Kinematics" and possibly "Quantum Logics."

***

Weyl's last book, *Symmetry*,[52] which he called (shortly before his death) his swan-song, stands in such a close connection with Group Theory and Quantum Mechanics that I may be allowed to add a few words about it. The small book sketches, in three short lectures, the role played by symmetry in mathematics, science, and the arts.

The book has been criticized in a recent issue of *The Mathematical Intelligencer* by Prof. Branco Grünbaum for its assertion that the old Egyptians had already found and displayed *all* 17 regular tessellations of the Euclidian plane.[53] He adds that Andreas Speiser, too, in his *Theorie der Gruppen*,[54] had been "beating around the bush" about this, but states that he, Gruenbaum, had never seen nor found a single ornament from this period that displays a triangular or a hexagonal symmetry, of which there are five. As I happen to have also made this statement in a paper of mine, I may be allowed to add a few words. I cannot, and do not want to, doubt Prof. Grünbaum's claim. With respect to my own article I am happy about Prof. Grünbaum's claim since it would prove the complete independence of the Mycenean hexagonal symmetries about which I had written.[55] But even then, notice that the dictum "the old Egyptians had already found *all* diagonal and *all* tetragonal symmetries" would still hold. I stress the word "all." It must especially be stressed that it is not at all true that the Egyptians found the trivial patterns while they missed the non-trivial ones. Quite the contrary: they found two highly non-trivial symmetries, those, incidentally, which Weyl in his first book, *Die Idee der*

[51] Paul A.M. Dirac, *The Principles of Quantum Mechanics*, 4th ed. (Oxford: Oxford University Press, 1958), p. ix.

[52] H. Weyl, *Symmetry* (Princeton: Princeton University Press, 1952).

[53] B. Grünbaum, The Emperor's New Clothes: Full Regalia, G string, or Nothing? *The Mathematical Intelligencer* 6, 4 (1984): 47–53.

[54] A. Speiser, *Theorie der Gruppen von endlicher Ordnung, op. cit.*

[55] See "The Symmetry of the Ornament on a Jewel of the Treasure of Mycenae" in this present volume, pp. 1–8.

Riemannschen Fläche, had called *eine Paddelbewegung*, a paddle motion.[56] Of the five which they missed, only one is really non-trivial. Perhaps we would do better to say that the predilection of the Egyptians for the tetragonal symmetries determined their style.

At any rate, they found *all* tetragonal symmetries, and therefore in a higher sense Weyl and Speiser were right, for the Egyptians could not have achieved this without *having thought* about symmetries *systematically*. This is what makes the occurance so interesting for the historian of science.

As Weyl says precisely in the introduction to *Gruppentheorie und Quantenmechanik*:

> While the quantum theory can be traced back only as far as 1900, the origin of the *theory of groups* is lost in a past scarcely accessible to history; the earliest works of art show that the symmetry groups of plane figures were even then already known ...[57]

I may add that the question of "mathematics and arts" preoccupied Andreas Speiser, for whom it became the pet topic of his philosophical research and reflection, more than it did Weyl. During his student years, Speiser had thought about and pursued these questions intensely and made it his preferred object of philosophical reflection, while Weyl's contributions to his relation in his whole work are rare. His interest was overwhelmingly directed towards the relation between mathematics and the sciences.

***

Three books were written which bore almost the same title, *Group theory and Quantum mechanics*: Weyl's, Wigner's and van der Waerden's. While it would be silly to classify them according to personal preferences, a comparison is interesting.

Van der Waerden's *Die Gruppentheoretische Methode in der Quantenmechanik*[58] is the most *compact* one; for someone who works uniquely in spectroscopy this is the shortest, albeit very steep, access to Quantum Mechanics. A special merit of this book is the formal spinor calculus invented by its author. Wigner's book, *Gruppentheorie und ihre Anwendung auf die Quantenmechanik der Atome*,[59] is the one of the three that found the greatest favour with physicists. I still remember the days during the summer holidays when I studied this book. To this day, the book remains the faithful companion of many theoreticians and experimentalists alike.

But it is fair to say – and this is surprising if you consider that Weyl was first of all a mathematician – that Weyl's book alone provides a broad and, in principle, complete introduction to the whole of Quantum Mechanics such as it was known at

---

[56] Weyl, *Die Idee der Riemannschen Fläche*, 3rd ed., p. 3, note 1; Engl. trans. *The Concept of a Riemann Surface*, p. 28, note 15.

[57] *Während man die Quantenmechanik höchstens bis zum Jahre 1900 zurückdatieren kann, verliert sich der Ursprung der 'Gruppentheorie' in die fernste, der Historie kaum mehr durchdringliche Vergangenheit. Die frühesten Kunstwerke belegen die Entdecktheit der Symmetriegruppen ebener Figuren* (Weyl, *Gruppentheorie und Quantenmechaniks*, p. 2; Engl. trans. *Theory of Groups and Quantum Mechanics*, p. xxi).

[58] B. L. van der Waerden, *Die Gruppentheoretische Methode in der Quantenmechanik* (Berlin: Springer, 1932), Engl. trans. *Group Theory and Quantum Mechanics* (Berlin: Springer, 1974).

[59] Eugene P. Wigner, *Gruppentheorie und ihre Anwendung auf die Quantenmechanik der Atomspektren* (Braunschweig: F. Vieweg & Sohn , 1931); Engl. trans. *Group theory and its Application to the Quantum Mechanics of Atomic Spectra*, J. J. Griffin, trans. (New York: Academic Press. 1959).

the time. His book is also the one that tackles problems of chemistry. As I made clear, it is (to paraphrase Weyl's teacher, Felix Klein) an "introduction to the fundamentals of Quantum Mechanics from a *higher* point of view."

Whatever we feel about the scientific merits of his book, the riches which we find in all its parts do not sufficiently explain the fascination which it has again and again exerted on its readers. I do not think that I was the only one who was fascinated by it, nor do I think that the number of those who, like me, were fascinated, was small.

Introduction

Allow me to recall that the young scientist formed at Caltech, who had won acclaim through a series of articles on special subjects, soon acquired extraordinary fame through his comprehensive *Handbuch* articles, the second of which was written with Walter Noll, where they reformulated the foundations of the mechanics of continua.[2] The mechanics of continua, which before had appeared in all textbooks as a conglomerate, even a sand hcap, of sometimes isolated subjects between which there was no connection, logical or mathematical, now became connected and unified by a powerful system of axioms as Hilbert had postulated at the beginning of the century.

This unification of the entire field of mechanics was made possible, among other things, through the sharp distinction between dynamical principles and constitutive equations. The latter formulates the special properties of the materials only, something that cannot be deduced in classical mechanics, but must be left to quantum mechanics. In classical mechanics such a property must be formulated by an additional hypothesis. Noll and Truesdell then showed that this distinction could be traced back to Cauchy and from Cauchy to Euler and to Jacob Bernoulli. Thereby we have now stepped into the domain of the history of science.

But here we must pause for a moment. While on the one hand it is obvious that having worked out a systematic organization of mechanics is indeed an incomparable preparation for analyzing, ordering and understanding the historical discoveries and the various processes of the development of science, on the other hand one cannot stress enough that science and history are two radically different endeavors of the human spirit.

The essence of science lies in its property of being systematic, since ultimately science always wishes to grasp the laws of nature, which it strives to uncover and to formulate in the simplest and most transparent form. But human history, and thus also the history of science, is the complete opposite of this: it is totally unsystematic, always complex and never simple nor transparent. Thus for writing the history of science two different – indeed, totally opposite – endeavors must simultaneously be at work in the same man; from the same man two almost irreconcilable gifts are

---

[1] This was a lecture presented at a symposium in honor of Clifford A. Truesdell held at the Scuola Normale Superiore in Pisa in 2002. Originally published in *Journal of Elasticity* **70** (2003): 39–53.

[2] See Clifford A. Truesdell and Richard Toupin, *Classical Field Theories of Mechanics, Handbuch der Physik*, vol. III/1 (Berlin: Springer-verlag, 1960) and Clifford A. Truesdell and Walter Noll, *Non-linear Field Theories of Mechanics, Handbuch der Physik*, vol. III/3 (Berlin: Springer-verlag, 1965; 2nd ed. 1992).

requested – gifts from his intellect as well as from his heart. This confrontation, one might say "clash," of the endeavor to systematize and to extract the universally valid from the documents which the historian finds before him, with the aim to determine the conditions under which a given, always unique, discovery was made, under very special circumstances and by one distinct individual different from all others, and then to interpret its significance for the development of science, is the character of the history of science. It is its very essence, even its unique prerogative, and also its characteristic charm.

Clifford A. Truesdell. Courtesy of the Archives of the Mathematisches Forschungsinstitut Oberwolfach

Above and beyond all of the interesting facts which we learn and insights which we gain about the progress that we are allowed to observe and learn to appreciate, it is precisely this constant confrontation which fascinates the reader of Truesdell's articles and books, and there perhaps more so than for any other historian of science I know. No wonder then that the interest of such a scientist-historian is directed especially towards the great systematizers and unifiers, those who at the end of previous long developments could lay the definite foundations of a whole field, like Newton, Euler, Cauchy, and Maxwell, or prepare new ways like Dirac did for quantum mechanics. But as we shall see, Truesdell also had a special admiration for Jacob Bernoulli. Here I shall restrict myself almost exclusively to what Truesdell did for the Euler Edition, which was the cause of my first contacts with him, and later, which is a very different story, for the Bernoulli Edition.

Truesdell's Introductions to Euler's Works on Hydrodynamics

Truesdell was invited in the 1950s by the mathematician Andreas Speiser, then General Editor of Euler's *Opera Omnia*, to edit volumes XII and XIII of the second series, which contain Euler's writings on hydrodynamics. The great systematizer that he was, as I need remind no one, Truesdell was much more cut out for that job than Speiser could have known. Indeed, Truesdell's introductions to volumes 12 and 13 of the second series[3] offer much more than what their titles promise and what Speiser could possibly have hoped for. He traced the subject back to its origins – Archimedes for the hydrostatic part and Torricelli for the hydrodynamic part – and then presented the development of these branches of science in detail, from the second half of the seventeenth century through the whole of the eighteenth century and beyond. Truesdell presented this scientific development, where many strands lived their separate lives for a long time, later became increasingly intertwined, and now form whole domains, but in the unplanned and never fully rationally explicable way that history, which does not care much for its future historians, always proceeds. Eventually he showed how these separate domains came together and why they had

---

[3] Clifford A. Truesdell, "Editor's Introduction: Rational Fluid Mechanics, 1687–1765," *Leonhardi Euleri Opera omnia*, Series II, vol. 12, pp. IX–CXXV (Zurich: Orell Füssli, 1954) and "Editor's Introduction," Series II, vol. 13, pp. VII–CXVII (Zurich: Orell Füssli, 1956).

to be formulated by a new mathematical language – a language whose very creation was stimulated largely by this development.

In these two introductions we can see how the systematic understanding of a science from today's point of view not only helps to unravel the various strands, but is indeed indispensable for showing exactly what each strand and each domain has contributed, and thus doing justice to each of them. Here it is not so much the experimental versus the theoretical progress that must be balanced, for, as Truesdell observed, because of the sudden enormous progress of mathematics during this period, theory was then mostly far ahead. (Incidentally, this is perhaps the secret reason why this period was always, and still is today, so neglected by the historians of physics.) Rather, the difficulty is to attribute proper credit not only to the stimuli that comes on the one hand from new, now suddenly accessible, problems of mechanics, and on the other from the impulses due to the analytical, and to a minor degree geometric, discoveries made around the middle of the century.

First we have Newton's heritage, deposited in the second book of the *Principia.* It is well known that precisely here are found some of the book's most difficult, even obscure, passages. Truesdell's summary of it and his evaluations are extremely succinct; they fill only two pages. Even so, they are a very valuable help to the reader. This is especially true for his careful separation of what Newton had derived by starting with the corpuscular view, from those results that he had derived using a continuum view, and finally, from those results for which he used both kinds of assumptions. The only regret the reader feels is that this whole section is so short.

Then we find Truesdell's examination of Daniel Bernoulli's *Hydrodynamica.* This book contains, albeit in splendid isolation from the rest, the celebrated excursion into the kinetic theory of gases, as we say today, which Truesdell places in the proper historical perspective. Here Daniel was over a century ahead of his time. But then in this book, hydrostatics is unified with hydraulics. This was the first of the two big unifications of theories during this period. It was achieved through the celebrated Bernoulli equation, which, as Truesdell carefully explains, was by no means written by Bernoulli in the form in which we know it today. Indeed, no fewer than four major discoveries were needed before its transparent form, due to Euler, became possible.

The first of these four discoveries was Johann I Bernoulli's introduction of Newton's concept of force into hydrodynamics, if only in a one-dimensional setting. It was Truesdell who rediscovered this fact, already acknowledged by Euler, thus clearing Johann from the reproach of having plagiarized his own son. Later, István Szabó confirmed Truesdell's vindication in even stronger terms. Truesdell also underlined the fact that the roots of Euler's hydrodynamics were to be found in Johann's rather than in Daniel's work.

The second important discovery was Euler's deeply penetrating understanding of what we call today an "inertial system" and, in particular, the fact that only in such a frame can one expect the laws of nature to have a simple form, a form that one can guess through mathematical or physical imagination. Truesdell characterized Euler's "Decouverte d'un nouveau principe de Mecanique" (E177) as one that would change the whole face of mechanics. This is the moment to mention the recent book by the Italian mathematician and historian Giulio Maltese in Rome, *La storia di "F = ma".* This book deals with the development of particle mechanics from Newton to Euler and it is perhaps the first one to profit fully from Truesdell's work.[4]

---

[4] G. Maltese, *La Storia di "F = ma". La seconda legge del moto nel XVIII secolo* (Florence: Olschki, 1992). After the meeting in memory of Clifford Truesdell in Pisa, there appeared a

The third important discovery was Euler's creation of the concept of "inner pressure," as opposed to Stevin's "outer pressure." This new concept was the key to his equations.

The fourth important discovery lies on the mathematical side of this development. It was the formulation of a field description due to d'Alembert, a discovery that was again first pointed out by Truesdell. On the mathematical side, one can, of course, directly observe the greatest step due to the mathematics of the eighteenth century: the systematic development of partial differential equations, again due largely to Euler and to d'Alembert in their work on hydrodynamics. The interplay between these two lines of research were masterfully pursued and presented by Truesdell. If we think how loath scientists usually are to recognize how much their own field owes to the progress of other areas, Truesdell's presentation is doubly remarkable and welcome.

Thus the ingredients were together for the equations of hydrodynamics, published in 1752 by Euler in his paper "Principia motus fluidorum" (Principles of the motion of fluids, E258). Thereby, hydrodynamics and aerodynamics were unified into one dynamical theory, which can be applied equally to either of them by selecting a special constitutive equation. These equations mark more than just a period in the history of science: they represent the first field theory in physics! The Bernoulli equation now appears in the form we know it today, namely as an integral of the equations of motion, valid only under certain circumstances.

To this culminating point in the history of mechanics there corresponds a culmination in Truesdell's narratives: his commentaries on Euler's two comprehensive presentations of hydrodynamics. The first of these consisted of three papers written in the 1750s in French (E225, E226 and E227), and the second was a series written in the 1760s in Latin (E258, E396, E409 and E424). The latter were part of Euler's great project, conceived in 1734, to present the entire science of mechanics in six volumes. The first of Euler's presentations reflects the moment of the discovery when it was still fresh in his mind, while the second, more polished, shows the ambition to present everything as clearly as possible, even for beginners.

I mention here only Truesdell's comments on Euler's paper, "De motu fluidorum a diverso caloris gradu oriundo" (On the motion of fluids arising from different degrees of heat, E331), written in 1764. This paper belongs to the prehistory of meteorology. Euler pursued a line taken up 150 years later by Milanković, but Truesdell pointed out, and here again he was the first to do this, what this paper brought to thermodynamics.

However, Truesdell by no means restricted his account to the leading figures of history; even though they receive the lion's share of attention, the others are not overlooked. Thus, the work of Simon Stevin, one of those who is always underestimated, eloquently receives his due. Truesdell also draws the reader's attention to Jacob Hermann, Jacob Bernoulli's most important disciple next to his brother Johann. It will be an important task of the Bernoulli Edition, to scrutinize Hermann's works for important contributions.

In the introduction to volume 13 one finds Truesdell's account of the development of acoustics. Besides Euler's, two names stand out here: Daniel Bernoulli and Lagrange. I prefer to mention Bernoulli's work later in connection with

---

second volume by Giulio Maltese, *Da "F = ma" alle leggi cardinali del moto* (Milan: Hoepli, 2001), a worthy successor of the first volume. One must hope that many readers and especially students will take profit from it. It is written in the spirit of Truesdell's pioneering work and it is a monument to it.

the introduction to the history of elasticity. Of Lagrange's work, Truesdell gave a detailed account of correspondence with Euler as well as of Lagrange's own published papers. He wrote that the velocity-potential theorem and the impulse theorem are first-rate creative works and Lagrange's greatest discoveries in fluid dynamics.

For organizing this vast amount of material, as one can appreciate, an intimate acquaintance with the sources is not enough. What was essential here was a deep insight into the science of mechanics itself – an insight that could only be gained from its most modern formulation. Such an understanding was essential for providing a reference system systematic and powerful enough to organize this enormous body of material adequately. It is the combination of both of these qualities, which Truesdell possessed to the highest degree, which lends dimension to his presentation and gives a dramatic quality to the narrative at the supreme moments.

Let us return to Euler's equations and their two great presentations. From Euler these equations came to Cauchy, who, with the help of his new creation – the stress tensor – laid the foundation of the theory of elasticity in its definite form. During the same period the field idea, through Euler's *Letters to a German Princess*, reached Faraday, who quoted them frequently in his journal, and from Cauchy and Faraday the concept of a field came to Maxwell. One may fairly say that either of Euler's great works – especially the Latin one, if supplemented with some of Truesdell's comments which place them in a modern perspective – is (or alas, as I rather must say, would still be) today the best introduction to hydrodynamics at the level of high school, and even more so at the level of university teaching. The presentation of the idea of what a law of nature is, how it works and what use is made of it under special circumstances, including technical applications, has hardly ever been surpassed. This presentation is never clouded by mere formalisms nor by going into the details of technological applications which, more often than not, just obscures the basic idea. By Euler and by Truesdell one is led from one essential point to the next with only the absolutely necessary excursions until one reaches the summit.

Truesdell's Introduction to Euler's Work on Elasticity

The Rational Mechanics of Flexible or Elastic Bodies,[5] as the title implies, reviews its subject from Galileo's *Discorsi* to Lagrange's *Mechanique Analitique*. In fact, however, it opens with a survey of the prehistory of the whole field of early Greek antiquity and also deals with the Middle Ages. Here we find such Truesdellian pearls as the following: "Duhem's great historical studies showed that the apparent darkness of mediaeval physics is but darkness of our knowledge of it" (p. 18). Indeed this was a call to fill an immense gap in science. I will mention here only one glaring hole: our ignorance of mediaeval technology. How, exactly, did the mediaeval architects and engineers proceed with the construction of their enormous cathedrals, especially of the towers, which surpass in height, refinement and daring everything that the Greeks and even the Romans had achieved? Truesdell's clarion call ought to remind all historians of science as well as of the arts that they should direct their attention much more to the Middle Ages than they have done so far. He explicitly mentions the elusive Jordanus de Nemore and notes:

> The only writing of value on deformable bodies that I have been able to see is the fourth book of Jordan de Nemore's *Theory of Weight* (13th century), and remarkable it is, Western in spirit, ambitious beyond

---

[5] Clifford A. Truesdell, *The Rational Mechanics of Flexible or Elastic Bodies, 1638–1788, Leonhardi Euleri Opera Omnia,* Series II, vol. 11b (Zurich: Orell Füssli, 1960).

anything in the Greek or Arab tradition. The seventeen propositions on fluid flow, resistance, fracture and elasticity are all original.[6]

The next section contains an evaluation of Leonardo's achievements.

The whole account fills the separate, 400-page volume 11b that contains the introduction to the two volumes 10 and 11! An immense achievement: theories, experiments, simple historical facts, etc., are not only enumerated but thoroughly "digested," that is, their scientific content is explained and imbedded in the historical development. Satisfying both requirements makes, of course, heavy demands on the writer. Add to this that the account is based not only on the published writings, but also on an enormous number of letters, for instance, on the quite complex epistolary exchanges between the Bernoullis and Euler, which Truesdell searched through. From these we can get an idea of the Herculean labor that he has gone through. And besides theories, we also find experiments discussed – experiments performed, among others, by Musschenbroek, Giordano Riccati and Chladni – and the account reaches far into technology and engineering.

Since the field of elasticity has a more complex structure than hydrodynamics, its history presents more isolated strands and subdomains. Thus, a lucid arrangement is much more difficult to attain here. In this present account I must restrict myself to a small part of Truesdell's accounts of elasticity proper and the notion of flexibility. I shall begin with the latter.

While the modern theories of flexibility began with Galileo's and Mersenne's investigations of the string, it was only the investigation of static problems like the hanging cord and the suspension bridge that led science to ask for mechanical explanations. In 1712, Brook Taylor computed the fundamental frequency of the string. While a bit later, as we learn from Truesdell, Daniel Bernoulli had the theory of the overtones in his hand, he did not publish it. Instead he studied the double-, triple-, and multiple-pendulum, and asked the following question, characteristic of his whole research: Under what condition is the oscillation stationary? His answer was that a pendulum with n masses has exactly n stationary oscillations, and he computed up to $n = 5$ its n frequencies as a function of the various masses and lengths. Turning next to the hanging cord, by going to the limit he proved that it has infinitely many stationary oscillations given by the zeros of a new transcendental function, today denoted as the Bessel function $J_0(x)$.

A little later Johann I Bernoulli formulated the problem of the motion of several particles attached to a string. Truesdell saw that in this field as well he was the first to use Newton's equation of motion.

Then in 1743 d'Alembert, who had studied Daniel Bernoulli's two papers, published, in the *Traité de Dynamique*, a partial differential equation for the hanging cord, the first partial differential equation in mechanics. Truesdell called this discovery "a turning point in the whole history of mechanics" (p. 192). Then, in 1746, in his paper "Recherches sur la courbe que forme une corde tendue mise en vibration," d'Alembert presented the differential equation of the string, this time together with the solution. A short time later Euler published the solution found in a different way.

With these discoveries began the last of the big scientific polemics of the eighteenth century. Truesdell was highly critical of the polemic itself, which he called "deplorable."

---

[6] See Clifford A. Truesdell, *The Rational Mechanics of Flexible or Elastic Bodies, 1638–1788*, *Leonhardi Euleri Opera Omnia*, Series II, vol. 11b (Zurich: Orell Füssli, 1960), p. 18.

C. TRUESDELL

THE RATIONAL MECHANICS

OF FLEXIBLE OR ELASTIC BODIES

1638—1788

INTRODUCTION TO

LEONHARDI EULERI OPERA OMNIA

VOL. X ET XI SERIEI SECUNDAE

AUCTORITATE ET IMPENSIS

SOCIETATIS SCIENTIARUM NATURALIUM HELVETICAE

TURICI MCMLX

VENDITIONI EXPONUNT

ORELL FÜSSLI TURICI

Title page, Truesdell's *Rational Mechanics of Flexible or Elastic Bodies 1638–1788*. Courtesy of the Biblioteca "G. Peano", Department of Mathematics, University of Turin

He commented that it "confirms the principle that ever the greatest quantity of paper is smeared over with the dullest matter" (p. 237). But this "great quantity" was searched through by him with great care, and the questions at issue were explained by him incisively.

In this triangular struggle both d'Alembert and Euler maintained that only the partial differential equation would yield all solutions. Their quarrel concerned the class of functions that are admissible as solutions. While d'Alembert tenaciously, but erroneously, maintained that only the functions today called analytic can serve as solutions of a mechanical problem, Euler admitted a much larger class, the class of piecewise smooth functions. Hereby he achieved, in Truesdell's words, "the greatest advance of scientific methodology in the whole century" (p. 248) because it contradicted the Leibniz postulate that in mechanics functions must be analytic, a postulate which, according to Truesdell, had not been contradicted by anyone, not even by Newton. Daniel Bernoulli, however, claimed that the trigonometric series would yield all solutions equally well. As he had overlooked the arbitrary phases, he was wrong, as we know today. The approach of his adversaries carried the day, but 250 years later we can see that Bernoulli had in fact solved the first finite and infinite eigenvalue problems, which now occupy, after the past 75 years, the center of quantum mechanics. Thus are the meanderings of the developments of science!

All these problems, however, were merely one-dimensional. Only one two-dimensional problem from the field of flexibility was solved during this period: in 1759 Euler discovered the equation of the drum.

The elaboration of the theory of elastic bodies moved during the same period on very different lines, especially with respect to the mathematics involved. Truesdell traced its modern development back to Beeckman and to Galileo. In the *Discorsi*, Galileo had asked: what is the proportion between the minimal weight P_l needed to break a beam simply by elongation and the minimal weight P_t needed to break it transversally when one of its ends is clamped into a wall?

Galileo's prediction was

$$P_l : P_t = (b : a)N$$

where a and b are the length and breadth of the beam, respectively, and N is a numerical factor. By assuming a special model he then could compute $N = 1/2$. Mariotte, who tried to confirm Galileo's predictions experimentally, found $N = 1/4$ rather than 1/2, and he wrote his results to Leibniz. Leibniz suggested, in 1684, that one should also take into account the energy needed to bend the beam before it breaks. Using Hooke's law he found $N = 1/3$. According to Truesdell this was the first computation which took account of the dilation of fibres.

A few years later in 1687 Jacob Bernoulli wrote to Leibniz asking for explanations concerning his new calculus; in the same letter he included the results of his own experiments, which in certain cases clearly disproved Hooke's law. Leibniz, because of his absence on a trip, replied three years later. He suggested that Bernoulli should determine the exact form of a beam that is bent by a weight. Bernoulli at once set out to work on this problem, and indeed, by using the principle of the balance of moments of forces and his own "golden theorem" – his formula for the radius of curvature –, he found the solution in the form of an integral which contained an arbitrary function depending upon the constitutive relation between the stress and the strain. Here, in 1692, Bernoulli recognized the distinction between dynamical principle and constitutive equation, for his experiments had convinced him that there was no universal law valid for all materials. In the simplest case, namely when Hooke's law is assumed, the solution is given by the famous elliptic integral, discovered and discussed by him in this paper. Of his theory of the bent beam Truesdell wrote, "the deepest and most difficult problem yet to be solved in mechanics, is his alone" (p. 96).

Truesdell pursued Bernoulli's later investigations of this problem, especially regarding the location of the neutral fibre. It ended, as Antoine Parent showed, with a failure. Truesdell remarked:

> To the ironies and disappointments which filled [his] life must be added that while he originated or assembled all the apparatus sufficient to put [his final equation] on firm ground, he failed to do so, failed because his attempt was on too grand a scale (p. 109).

In fact even today the problem of the neutral fibre seems not to have been solved in full generality.

Truesdell's evaluation of Jacob Bernoulli's achievements is, "In our epoch for study, 1638-1788, but one other, Euler, is to build himself a like monument in our subject" (p. 109). On the other hand, it is characteristic of Truesdell that he devoted a full section to Parent, whom he rescues from near oblivion by showing that "Parent was the first to apply statical principles correctly to the tensions of the fibres of a beam, and that he recognized the existence of shearing stress" (p. 114). These are no small merits, indeed!

The next step in this development was taken by Daniel Bernoulli. He knew that Euler was working on a book on variational calculus, and suggested to him a minimum principle for the potential elastic energy stored in a curved beam. Euler immediately worked out its consequences which he annexed as the first appendix to his *Methodus Inveniendi Lineas Curvas*, where he derived a multitude of new results.

Truesdell also commented on Euler's further work in this field, for example, his discovery of the shear force, and Coulomb's discovery of the shear stress. In this field all correctly solved problems were, again, one-dimensional. Even the problem of the oscillations of a massive plate was missed by Jacob II Bernoulli, although, as Diarmuid Mathúna showed, only by a hair's breadth![7] Here Truesdell was a bit severe, for the definitive solution would only be given by Lagrange more than thirty years later.

Especially impressive is Truesdell's "modern evaluation," which fills the last ten pages of the book. He divides the task into three parts: the evaluation of Analysis,

---

[7] See Diarmuid Ó. Mathúna, "Jacob II Bernoulli and the Problem of the Vibrating Plate," pp. 165–177 in *Entre mécanique et architecture / Between Mechanics and Architecture*, Edoardo Benvenuto and Patricia Radelet-de Grave, eds. (Basel: Birkhäuser, 1995).

Geometry and Mechanics. Who else could have dared to evaluate three so basically different histories? I suspect that even a careful and informed reader will discover in these few pages, here and there, a fundamental aspect of something he believed to have well understood but had in fact failed to grasp. I repeat here the first sentences of the first summary. The triumphant lines show Truesdell's rightful pride in his own beloved science, Rational Mechanics. He writes:

> Prior to 1730, researches on continuum mechanics applied mathematical techniques already developed in other subjects, notably in geometry and in the mechanics of point masses. Starting with the research on vibrating systems by Daniel Bernoulli and Euler, the situation was completely inverted. From then on until the end of the century, *continuum mechanics gave rise to all the major new problems of analysis*" (p. 416).

On the two last pages of the book Truesdell asked why the foundations to a complete theory of elasticity escaped this period, and writes:

> Neither physical intuition nor experiment was what was needed here; rather, as both Euler and Chladni said, it was *want of differential geometry* that blocked the way to theories of deformable surfaces and solids" (p. 427).

Finally, after mentioning that Euler had introduced all elements of the strain tensor in a paper on hydrodynamics, he notes on the very last page:

> In surveying all these brilliant individual achievements ..., we are driven to ask why, when Euler had succeeded in 1752 in creating a general theory of perfect fluids ..., nevertheless after many more years he failed to reach a general theory of elasticity (p. 428).

His answer was:

> To succeed in hydrodynamics, the only hope lay in abandoning a one-dimensional approach. But for elastic or flexible bodies one-dimensional theories led to one triumph after another. It was the *brilliant successes* of the special theories that blocked the way to the general theory, for nothing is harder to surmount than a corpus of true but too special knowledge" (p. 428).

Here I could give but an insufficient account of this monumental work of Truesdell: the history of the theory of elasticity is now, probably due to him, the best charted and the best investigated domain of the history of physics. I remain convinced that the three introductions to which I have referred are the best guide to a deeper understanding and further study of the history of classical mechanics and indeed of the history of science; every time I open one of them I find something new and interesting that had escaped me.

The Concepts and Logic of Classical Thermodynamics as a Theory of Heat Engines. Rigorously Constructed upon the Foundation Laid by S. Carnot and F. Reech

Before turning to my second main topic, Truesdell's work for the Bernoulli Edition, I would like to mention his book with Subramanyam Bharatha.[8] Even if it lies somewhat apart from the other works of this account, it brings to the fore several characteristics of Truesdell that seem to me fundamental for his thinking as well as significant. I quoted the full title of the book, since this book brings together, like no other books of his that I know, science, history and, not surprisingly, conceptual logic. And in no other book of those which I know, is Truesdell so preoccupied with teaching. Not that I would recommend the book as a textbook for students, but I recommend it highly to all those who teach physics. The aim of the book is to construct a rigorous foundation of classical thermodynamics based on the idea of the Carnot cycle. Here "rigorous" refers not only to mathematical rigor, but also, and in fact even more so, to conceptual rigor: to a clear and adequate introduction and a sharp definition of all concepts that will be used in the equations as well as a precise reference as to how they are to be measured, i.e., how they are connected with experiment. The very first sentence of the preface makes this clear and is, at the same time, a "critique," in the sense of Kant, of the possibility of writing the history of science. He writes, "I do not think it possible to write the history of a science until that science itself shall have been understood, thanks to a clear, explicit, and decent logical structure" (p. vii). I have hardly found in Truesdell's work positive references to philosophy, and probably he would be surprised to be referred to as a philosopher; yet what he notes here and in the rest of the section is as important a contribution to the philosophy of science and to the philosophy of history of science as I have ever heard. Perhaps Truesdell, were he here, would react to this compliment with a little smile. On the other hand, the aim which he pursued as a historian is expressed by his dedication of the book "as an expression of respectful gratitude for the legacy of the great French thermodynamicists Carnot, Reech, Duhem" (p. iv). Incidentally, this dedication refutes the accusation that I sometimes heard that his was an anti-French bias. In three sections "Calorimetry," "Carnot's General Axiom," and "Universal Efficiency of Ordinary Carnot Cycles," the results are presented.

I believe that the progress of science consists in establishing connections between various phenomena, and between phenomena and their measurements, and includes a process called the formation of theories, that is, constructing connections between different restricted theories by erecting greater and increasingly comprehensive theories. The significance of the book may then perhaps lie in the first place that no other book of those that I have consulted connects thermodynamics so cogently and intimately to classical mechanics.

But the book has another distinction too. It is a book written especially for the teacher, I dare say even for the teacher, who must speak to beginners. In other words, the book also has a pedagogical aim. If it is not directly a textbook, this is only because the authors wished to prove that their approach is powerful enough for coming to grips with all situations that the practical applications demand. Hence the careful analytical generalizations to cases where functions that are only piecewise smooth are needed, etc. But for the explanation of the thermodynamical principles themselves, these technical details are not necessary and can easily be suppressed by

---

[8] Clifford A. Truesdell and Subramanyam Bharatha, *The Concepts and Logic of Classical Thermodynamics as a Theory of Heat Engines. Rigorously Constructed upon the Foundation Laid by S. Carnot and F. Reech* (New York: Springer, 1977).

the teacher. But above all what the teacher can learn and teach is not so much the mathematical rigor, but the conceptual rigor of the theory or of any theory for that matter, and the importance of a careful introduction and explanation of all concepts. The importance, for instance, of the eternal question that looms over the beginning of all introductory courses on mechanics: "what exactly is the force, professor?"

When I think of my own lecturing, my greatest regret is that I concentrated too little and too late on the careful introduction of all concepts used in physics, and that I spent in my lectures too little time on their discussion. It is with the help of sentences that we prescribe the setting of a reproducible experiment – the concepts connect the experiments with the mathematical formalisms.

Another fundamental point made clear in this book is that all theories are always valid only with respect to a certain domain of the variables and under certain restrictions. It is the neglect of these caveats which makes possible only pseudo-philosophical and pseudo-scientific generalizations. Here teachers can learn much that will prevent a certain boastful offer of their merchandise and at the same time make the understanding of what they present easier since it is focussed. I learned myself, for example, that the teacher must immediately, at the beginning of each course, enumerate explicitly all restrictions only under which the predictions of the theory are valid.

Over twenty years ago I had invited Truesdell to Louvain-la-Neuve to give a series of lectures. In the first one he outlined the content of this book in one hour, overestimating, of course, his audience, which was oriented mainly towards quantum mechanics and its applications. Then he changed to other subjects. But he presented me with a copy of the book, and when I had read it, I regretted deeply that the whole series of lectures was not directed to this one topic. I did not dare say it to him. But later, over lunch, I told him that I was particularly impressed by one special topic, namely, his treatment of the anomaly of water between 0 and 4 degrees, of which I never had seen an adequate presentation. He then said approvingly, "I can tell you that this subject was my special goal for writing this booklet," and then taking his glass he invited me henceforth to call him "Clifford," which from here on I shall also do in this discourse!

During his stay in Louvain-la-Neuve there was one other topic towards which many of our conversations were directed again and again. This was the Bernoulli Edition, to which I shall now turn.

Truesdell's contribution to the restart of the Bernoulli Edition

This other topic which occupied Clifford and me was precisely the new beginning of the Bernoulli Edition, and here I must now go back a few years. Clifford had become involved with the Euler Edition, as I mentioned earlier, through the mathematician Andreas Speiser, an uncle of mine, with whom I had close contacts. My uncle was extremely proud of this acquisition and spoke to me often and enthusiastically about Clifford. He gave me *separata* of the two introductions to the hydrodynamical works, and my uncle's enthusiasm also caught on with me.

I made Clifford's acquaintance in 1957 on the occasion of Euler's 250th anniversary, where at my uncle's invitation he was the main speaker at the official university ceremony. A few years later he wrote me a complimentary letter on my own introduction to Euler's works in the domain of physical optics. Meanwhile I had begun to read his introductions, so that when J. O. Fleckenstein asked me to succeed Hans Straub as the editor of the works of Daniel Bernoulli, I said "Yes, but ...." Namely, I stated as a condition that I should be paid the equivalent of a part-time assistant. It so happened that a few months earlier a young student had asked me if she could write a Ph.D. thesis under my direction. I wished to accept her, for she had definitely *une tête bien organisée*, but no post seemed free. Hence my proposal to Fleckenstein. I succeeded in persuading the student to work on the history of science, although at first she found this a puzzling and somewhat dubious proposition, and Fleckenstein arranged the financial side. Today this young student, Patricia Radelet-de Grave, is Professor at the Universite Catholique de Louvain, where she teaches the history of science. She has now succeeded me as general editor of the Bernoulli Edition, while Dr. Bruna Gaino serves as the Edition's secretary.

Studying Clifford's introductions, I had become convinced that he was the best possible guide and counsellor for launching the whole enterprise again. In 1975, I was in New York when I received a call from Clifford, who inquired about what was going on in the Bernoulli Edition. I gave him the little information I had, but only later did I find out that he had not been terribly excited by my answers. Nevertheless, later, as I shall mention, he accepted an invitation to be the editor of Daniel Bernoulli's work on hydrodynamics. To my surprise he wrote to me that he was not satisfied with his earlier work in the Euler Edition. His, as I noted before, was the only critical voice about these introductions, of which I ever heard.

In 1980, when I succeeded Fleckenstein as the Editor of the whole Edition, he encouraged me to ask André Weil to become an editor of the works of Jacob Bernoulli. Weil very kindly accepted first the volume on Analysis and then later also the volume on Differential Geometry. Also, from Weil I learned much about the art of editing, and I wish to state here too that I cannot remember one difficult moment with him. A bit later Weil introduced me to Herman Goldstine, who edited the volume on Variational Calculus containing works of Jacob and of Johann. Meanwhile Dr. Radelet and I produced plans for editing the works of all Bernoullis; so far there had not been any plans nor any reliable estimates of the work at all, the old ones all being much too low. All these questions and many more, including the choice of the typography and the design of the volumes, were then discussed with Clifford in Louvain-la-Neuve. He took a detailed interest in all problems, and Dr. Radelet and I learned much from him.

At last we could publish the plans, printed in 1982 in an illustrated brochure, which also contained a presentation of the Bernoulli family and the importance of each member. The plans consisted of:

i. a presentation of the whole project, including what had already been achieved and the distribution of the works into volumes;
ii. a determination of our priorities: first to complete what was begun, i.e., Jacob Bernoulli and Daniel Bernoulli and three volumes with the letters of Johann I Bernoulli, and only then to complete in a second stage all works of Johann I and the "minor" Bernoulli's (this stage has now begun with the 2008 publication of the works of Johann I and Nicolaus II edited by Piero Villaggio from Pisa). A third stage was foreseen for the letters; for their publication a project was worked out by Fritz Nagel and myself;
iii. a list with the names of all editors of the first stage.

At first, when I wrote to Clifford about this brochure, he did not seem terribly impressed, but when I sent him two copies he sent me his enthusiastic congratulations, together with a list of friends and colleagues to whom he invited me to send a copy.

These plans, with the appointment of editors for the first stage, were the basis of the 1982 restart of the Bernoulli Edition. The same year, on the occasion of the bicentenary of Daniel Bernoulli's death, the Curatorium, under its president the historian A. Gasser, organized a Symposium. The main speaker was Clifford, who in the *Alte Aula*, between the portraits of Daniel Bernoulli and Euler, gave a speech about the research of both on the theory of oscillations, evaluating the strengths and the weaknesses of both of them.

This was the beginning of a series of exchanges concerning the progress of the edition, and especially of my requests of Clifford's opinion on various questions. At the Symposium we presented the first new volume edited by L. P. Bouckaert and B. L. van der Waerden. It received, besides several favorable reviews, one that put the Edition at its start into serious trouble. Again Clifford came to our rescue and, in a letter to the Swiss National Science Foundation (SNScF), refuted all points of the review, with the exception of one. His letter especially restored, as I was told, the confidence of the SNScF. Incidentally, the author of the review later graciously apologized to Dr. Radelet.

Of course, we had all very much hoped that Clifford might deposit the two volumes containing Daniel Bernoulli's work on hydrodynamics. This was not to be. As all know, an enormous load of work kept him busy beyond his forces, and his health eventually failed him, although he was in the best of care. He had to resign from his engagement, and he advised me to invite the Russian Academician Gleb Mikhailov to take over.

But, thanks to his friend Andre Weil, Clifford nevertheless became a Bernoulli Editor! Indeed Weil had advised me to edit completely all letters of Jacob Bernoulli's correspondence with Leibniz and I had unhesitatingly accepted his advice, unaware of the existence of an agreement between the Leibniz and the Bernoulli Edition, which had left the editing of the letters with Leibniz to her sister-edition in Hanover; these letters were, of course, the most interesting ones. But when I explained the situation to our colleagues of the Leibniz Edition, they very generously accepted our plans, provided we would not undertake a "critical edition." During his work on the edition Weil persuaded Clifford to write an introduction to the parts of the correspondence that dealt with questions of mechanics.[9] It is there that Leibniz drew Bernoulli's attention to the problem of the curved arc. The result of Weil's invitation was again the appearance of a very penetrating introduction.

Thus the Edition is fortunate that Clifford's name will remain connected to it, and especially as an editor of Jacob Bernoulli, for whom he had done so much. But even without this turn of events, after thirty years of work for the Bernoulli Edition, I can firmly state that no one has done more to make the new beginning of the Edition in 1982 possible than Clifford Truesdell.

Truesdell the Artist and the Man

My report on Truesdell's work for the Euler and the Bernoulli Edition must end here, although I could go on at length. But there remains a question: neither Clifford's scientific expertise nor his penetration into history alone can explain the full fascination which his works exert on their reader. We know that all scientific theories are even at best only approximations of the observed world and the same, but even more so, holds for the reconstruction of the historical path on which they were founded. Clifford, more than most historians, was aware of this. All too often we must be satisfied with guesses. Thus, like the scientific theories, the historical reconstruction to some extent always remains a construct. So, what then produces the great satisfaction that we experience when we read Truesdell's works? To discover

---

[9] Clifford A. Truesdell, "Mechanics, especially elasticity, in the Correspondence of Jacob Bernoulli with Leibniz." Pp. 13–26 in *Der Briefwechsel von Jacob Bernoulli* André Weil, ed. *Die Werke von Jakob Bernoulli*, vol. 4, no. 2 (Basel: Birkhäuser, 1993).

the answer we must, I think, turn to another field, enter another dimension: the realm of beauty. This man, who was so attached to all arts – music, painting, old books, etc., – was himself an artist. He has composed his books, in the double sense of this word. As much as his search for scientific precision in all details would allow it, his books are beautifully constructed!

This brings me necessarily to Clifford the private man. Everyone who had the intense pleasure of being received in the Palazzetto knows what I mean when I speak of the carefully chosen objects of the collection, their carefully thought out presentation, and especially their owners' passionate interests in all arts and also in the history of the arts. Here too one could experience the truly enlightening comments which their guests received. One could watch how they stimulated through their interest in all crafts the artisans of Baltimore who made the Palazzetto into what it became. As you realize, I slipped now, almost unconsciously, into the plural: the treasures of the Palazzetto were offered to its guests by a couple! And so this was more than only their home! Can we imagine Clifford's tremendous outpourings without the constant and intense help of Charlotte? Can we imagine this without her painstaking proofreading of his books, her corrections, her meticulous improvements of the last details, her conscientious organization of the Archive as well as the classification of Clifford's correspondence in a private archive? Of course we cannot. Just as the Palazzetto's hospitality was the work of both, the Palazzetto's soul was Clifford and Charlotte. And Charlotte made sure that in spite of his harsh afflictions Clifford could spend his last years there in dignity. For this, all of Clifford's friends will always be in her debt, and they will remain grateful to Charlotte and Clifford.

Acknowledgments

It is a pleasure to thank Professor G. Capriz for the invitation to contribute to the Pisa Meeting in memory of Clifford Truesdell and to Professors Chi-Sing Man and R. Fosdick for their careful editing of this article. Also, I would like to thank Professors Luigi A. Radicati di Brozolo and, especially, Piero Villaggio for many interesting conversations, and my wife for linguistic advice.

Publishing Complete Works of the Great Scientists: An International Undertaking[1]

David Speiser and Patricia Radelet-de Grave

Introduction

The importance of publishing complete editions of the works of the great scientists of the past may be perceived more clearly through comparison of the world's two great heritages: the artistic and the scientific. Everyone is in direct contact with the artistic heritage: we can all visit churches and at least see from the outside the great palaces and castles. Most sculptures and paintings are accessible through museums. Musical works are kept alive through the many music schools and opera houses and may even be heard at home thanks to the radio and recordings. Bookshops and libraries give us easy access to literary works of all countries and all periods.

This situation is in singular contrast with that relating to our scientific heritage. It is generally very difficult, even under the best conditions, to gain access to the great discoveries that constitute the masterpieces of our scientific past. They are buried in libraries, lost in books and journals, in many cases hidden in manuscripts. What is more, even if one of these texts is found, it is very difficult to read, perhaps obscure.

Though educated people today may be keenly interested in the great discoveries of the past, they are usually obliged to learn about them in popular works whose authors only too often know the originals at second, third or even fourth hand.

Yet no one would deny the profound influence of science on our thinking, still less the extent to which technology permeates our lives. Their influence is so great that one is led to ask how science and technology, which always go hand in hand, have become what they are today. The answer, always colored by philosophical, political or other interests, is frequently in contradiction with the historical facts. In science books footnotes intended to refer to such and such a discovery are all too often incorrect, and many a theorem does not bear the name of the person who actually discovered it first: hence the need to go back to the original texts and make early studies accessible to people today.

Making scientific research accessible

What exactly is meant by "making scientific research accessible"? Newton's *Principia*, published in 1687, is probably the most famous work in the history of science. Many people imagine that this book, which is about 500 pages in length, contains the basic principles of the whole of modern mechanics as taught in the

---

[1] Originally published in English and French in *UNESCO: Impact of science on society* 160 (1990), 321–348. Particular thanks to Patricia Radelet-de Grave for permission to republish this work.

universities today. Many people also imagine that this book was read, studied and understood by all physicists in Newton's time, and by a large number in later times.

These beliefs are quite wrong. In fact, it would probably not be an exaggeration to say that only about a dozen scholars read and understood the work during the fifty years following its publication, and that few have studied it during two and a half centuries that have since elapsed. It is not that the book lacked significance. The reason is the extreme difficulty of the subject matter and the obscurity of the language: perhaps, too, the inadequacy – not always unintentional – of some of Newton's explanations. It is not surprising therefore that the descriptions of the contents of this book are often full of errors. To take just three examples:

- Newton confines his study to the systems of point-masses, deals very little with rigid bodies and not at all with elastic bodies;
- The famous equations attributed to Newton are not contained in the book in the form in which we know them today. It was only in 1750 that they were expressed in that form by Euler;
- The work contains very few infinitesimal formulations. It is therefore not sufficient to give even a physicist or an engineer of our time a copy of the *Principia*. The text would remain inaccessible, most of the chapters requiring detailed annotation.

Furthermore, the reader might be interested in purely historical questions such as: What was known before Newton? What new knowledge did he add? How and through what research were Newton's assertions transformed so that they could be formulated as they are today? What knowledge was added by his successors?

To reply to these questions, a detailed introduction and careful annotation would be required. (Many answers have, in fact, already been provided by D. T. Whiteside in his editing of the works of Newton.)

What we have just said concerning this particularly famous book is *a fortiori* true for many others, perhaps less well known, but in fact just as, or almost as, important, and often just as difficult. Even in the case of an author as lucid and clear as Euler, notes are essential to guide the reader through his labyrinthine work, of which the complete edition now totals seventy-two volumes.

What follows is an attempt to explain the importance of editions of complete works and answer questions such as: What is their special role? How is such an edition produced? The need for international collaboration will be stressed, and a few of the most important editions will be mentioned. In conclusion the relations between science history and science teaching and between complete editions and non-European countries will be considered.

A brief survey of a few editions of complete works

In all ages complete works have been published. According to the philosophers, we possess all the works of Plato—and this only because they were put together systematically. In 1744, Johann I Bernoulli published a complete edition of his own works in four volumes, introduced by a quatrain attributed to Voltaire himself. Four years later an edition of the complete works of his elder brother Jacob, who had died forty years earlier, was published, thanks to the nephew of Johann and Jacob, Nicolaus I Bernoulli, and the mathematician Gabriel Cramer. Such editions were expensive and often paid for by subscription, particularly in England. They have been invaluable and although they no longer meet present critical demands, they are still appreciated.

When education was organized at all levels within a public education system, states also began to use the system to promote this type of edition as a tribute to their great men and to enhance national prestige. In this way a number of editions of complete works were published, including those of Galileo in Italy, Descartes, Lagrange and Laplace in France, and Gauss in Germany. Such editions were often prepared by distinguished scholars: for example, the works of Descartes were edited by P. Tannery, those of Fourier by G. Darboux, and those of Newton, as we mentioned earlier, by D. T. Whiteside.

The smaller countries of Europe of course tended to lag behind, but their contribution was also important. We are in possession today of an edition of the complete works of Huygens, which the science historian A. Koyré described as "beyond all praise." It was produced by the Academy of Sciences of the Netherlands, which spared neither human nor financial resources. The result of an immense amount of work, it includes a large number of records and tables which are of considerable assistance to the researcher. It may be regretted that each volume does not contain a plan of the whole edition. In fact, since most libraries will only lend two or three volumes at a time out of the total of twenty-two, the reader may have difficulty in locating the topic studied. This criticism is made only to show that there is room for improvement in the art of publishing. It might also be noted that it would be inconceivable today for the members of the famous Academy to work anonymously. However that may be, this edition has always served as a reference for those that followed.

A plan for an edition of the complete works of Euler, probably the largest scientific work ever produced, was prepared by three German mathematicians, Ferdinand Rudio, Adolf Krazer and Paul Stäckel, before the first World War. The edition was based on a catalogue of all Euler's works (some 800 titles), which had been prepared shortly before by the Swede Gustav Eneström. The original plan ran to some fifty-four volumes. The first volume, for which the famous mathematician Heinrich Weber assumed responsibility, was published just before the war, in 1913. Later, it was realized that the original plan was quite inadequate and a new plan comprising seventy-four volumes was drawn up by the Swiss mathematician, Andreas Speiser.

Three features of this edition are worth mentioning:

- Geographically the edition was based in Switzerland, but it benefited from international collaboration, with scholars from at least six countries participating. It is clearly not by chance that the tradition of basing editions on international collaboration was born in a small country. Some years ago the initial plan was extended to include Euler's letters and manuscripts. This new part of the edition, prepared by Switzerland in collaboration with the USSR, will consist of between twelve and twenty volumes.
- In this part – and this is another new feature as compared with the very first volumes – Euler's works will be annotated and accompanied by an introduction. This critical apparatus will of course increase the value of the edition considerably, for it makes the writings and the work as a whole more accessible.
- Certain texts by Euler have also been accompanied by texts of other authors, to whose arguments he was replying.

The edition of the works of Leibniz is a special case. The total volume of his scientific and philosophical works exceeds that of the solely scientific work of Euler. In addition, the vast quantity of Leibniz's manuscript notes still extant raises infinitely complex problems.

In 1935, the Swiss mathematician Otto Spiess began to work on the Bernoulli edition. The edition was reorganized in 1982 and the work continues today with the collaboration of scholars from at least eight different countries. Among them are scientists and historians as eminent as B. L. van der Waerden, André Weil, Clifford A. Truesdell, Herman Goldstine and Gleb Mikhailov, all of whom consider that the work of a science historian deserves all their efforts. Of the forty-five or so volumes of which the edition will be composed, seven have been published and four are now in press or in preparation.[2]

This is not all, however. In many countries today a growing interest in such editions may be observed. France has decided to pay tribute not only to A. Clairaut and J. d'Alembert, but also to Monge and Desargues. Italy is preparing a number of editions, and in Belgium it has been decided to publish the works of G. Mercator.

The time required to complete such editions should not come as a surprise. For example, the edition of the works of Cauchy took over fifty years to complete. This makes it particularly difficult to organize the work and maintain continuity.

Why publish complete works?

Much of what has been said so far applies to the publication of any scientific text. What aspects are specific to complete editions?

One aspect that must be mentioned is the overwhelming importance and interconnection of the works of a relatively small number of scholars whom we describe as "great." It is these scholars who have often guided the others and opened new horizons for them. They have generally done this by making a synthesis of what is already known in a certain field and supplementing it with their personal findings, thus opening up new avenues of research. One has only to mention Copernicus, Kepler, Galileo, Newton and Euler.

Another aspect specific to complete editions is that the work of the science historian is greatly facilitated thereby. All too often research papers or articles are studied in isolation or in relation to too few other texts. Sight is thus lost of their special significance in the general development of the history of science. Theoretically, all or at least large parts of what has already been published should be republished and worked on again, so as to provide an overall view. Yet this is out of the question and almost absurd.

Complete editions therefore have the great advantage of presenting us with a full series of studies through which we can follow developments in a given field, the emergence and elaboration of the concepts relating to that field and the gradual progress in understanding, sometimes right up to complete understanding. It often suffices to compare this panorama with that to be gained from a study of the complete works of two or three other authors in order to acquire an overall view. One may see more clearly in this context the importance of including in the complete works writings which stimulated the thought of the author in question or which were a direct consequence of the published work. Thus the first volume of the complete works of Euler includes the famous marginal notes by Lagrange and another volume contains a reference to the work of the Italian Lorenzo Mascheroni. One has to acknowledge that these editions are not only essential instruments for the science historian but in fact form the very basis of the history of science.

---

[2] Editor's note. As of this present volume, seventeen volumes are available. I thank Patricia Radelet-de Grave for this information.

Who should and can work on these editions?

It seems clear that the responsibility for such joint editions devolves first on the scholars of the author's native country. It is there that most of the manuscripts are generally to be found and it is accordingly there too that the essential financial assistance is likely to be obtained. Nevertheless, as already pointed out, international collaboration is necessary, for science is carried on today on a world scale and this should apply also to the history of science.

Actually, the foundations of modern science were laid in a small number of western European countries. However, these early scientists based their research on the work of the Greeks and the Arabs and, through those two peoples, on that of the Egyptians, Babylonians, Sumerians, Indians, Chinese and others. The development of science is therefore a world phenomenon which it would hardly be appropriate to treat at a regional or even a continental level.

Another reason for associating all continents in this work is best explained once again by a parallel with art. Erwin Panofsky, the great art historian who emigrated to the United States in 1933, used to say that the position of the USA opened up possibilities for the art historian of which he never dreamed before. In Europe one was always restricted to national confines, which often distorted one's views. But from America one had a totally unprejudiced view of European art. The great distance meant that one saw European art as a vast panorama.

This is equally true of the history of science. For instance, at least four of the scientists to whom the American C. A. Truesdell gave due credit came from small countries, namely: Simon Stevin, Belgium; Christiaan Huygens, the Netherlands; Jacob Bernoulli and Leonhard Euler, Switzerland. Other examples are to be found in the work by A. Weil, which presents in Olympian fashion the discoveries made in the field of number theory by – in chronological order – a Greek, Diophantus of Alexandria; a Frenchman, Pierre de Fermat; a Swiss, Leonhard Euler; a Savoyard, Joseph Louis de Lagrange and another Frenchman, Adrien Marie Legendre.

So let us hope that these editions will find collaborators in all countries and all continents and particularly in what is now called the Third World. All those who have contributed to this work have profited from it, and for those coming from another continent the benefit will also be incalculable.

How is a complete edition produced?

Such a vast undertaking can only be carried out in stages: the first stage is the most difficult because it is the least gratifying. It is also generally the longest. It consists of collecting everything that the author has written, even if it is intended not to publish everything but to limit the edition to one or other of the following four types of material:

- published texts, namely, books and papers issued separately, or articles published in journals;
- manuscripts written with a view to the preparation of an article;
- diaries and manuscript notes;
- letters.

Clearly, when it is decided to publish only part of a work, the choice falls generally on the first and the last of these types of material. However, in order to make a judicious choice, it is essential to have seen the whole of the author's production. Unfortunately, one can never be sure of having all the material at hand. This is understandable in the case of letters, but even whole works are often difficult to track down. There comes a point at which the quest is no longer profitable and it is necessary to call a halt, at least provisionally, and to decide to pass on to the next stage.

The second stage consists of making a plan of the edition. A way of classifying the texts must be selected, then a way of distributing them in volumes. At this stage, too, the number of volumes and their size must be estimated. It should be noted that the principles of classification and distribution are different for each of the types of material mentioned. It should also be stressed that these principles may vary from one author to another and that the few ideas put forward here are given by way of example.

Published texts are often better rearranged according to subject matter – mainly because the reader is generally interested in research on a particular subject and this arrangement makes it easier to locate that subject in an often vast work. Then, however, the question that arises is whether the subject matter should be divided up according to current criteria, or according to those of the author's time. The first course is generally more convenient for the present-day reader, but it must be borne in mind that the subdivision of science into different branches has not always been the same and that it is constantly changing.

For more distant periods it is difficult or even impossible to choose between two sets of criteria; one is forced to make do with the classification of the period.

In any case the principles selected must be made clear to the reader in the introduction to the first volume of the edition and also in the first volume of each series. It is essential that the reader have ready access at all times to a general plan which will serve as a guide for an edition consisting of even twenty, let alone eighty-five volumes. This plan may often be set out in a one-page summary appearing in each volume. After all, these editions are not only for general reading; they are working tools, which must therefore be functional and enable the researcher to find at once the text sought for analysis. (In fact a classification according to current categories is almost inevitable – for reasons which will again be stressed.)

To return to the classification of the different types of material, within each group of texts concerning one subject, the chronological order almost automatically becomes imperative. However, it must be decided in each case to what extent categories should be subdivided and where the chronological classification should begin. Compromises will be inevitable. With manuscripts and diaries, it all depends on their size and on their relation to the other material.

With correspondence, on the other hand, the criteria are quite different. As several subjects may be dealt with in the same letter, classification by subjects is impossible and the chronological order has to be followed.

If justified by their number, the letters exchanged between two persons can of course be grouped together in a single subcategory. Within that subcategory, nevertheless, a chronological arrangement is customary. In any case it is essential to publish the letters of both correspondents together. If the number of correspondents is limited, the letters may be put together and the whole series therefore merely reproduced in chronological order.

The foregoing considerations concerning the organization of the edition would be incomplete if the question of which part of the work should be published first were not posed. The views of scientists consistently diverge with those of historians. The former are interested mainly in the published texts because these have determined the development of science, whereas the latter prefer to begin with the manuscripts and letters so as to follow the emergence of ideas.

The authors of this article are strongly in favor of the former solution, for the reason expressed by Jean Dieudonné: It is in the published texts that one normally finds the most maturely considered and clearest formulations. Moreover, if these texts are published first, they will be of use to the editors of the manuscript parts of the work. As to the deep understanding of an author's work and intellectual development, experience shows that this is reached only after long and arduous efforts on the part of several researchers.

Professional historians or philologists often criticize those who published complete works of scientists for having begun their task at the wrong end. They want the publication of the texts to be preceded by that of all the notes and papers that led up to the final published work. In the case of papers written before the invention of printing, they consider that one should begin by establishing the exact filiation of the manuscripts, as philologists have always done when editing the classic works of literature. In advocating this, they forget that we know these classics, which have been read and reread over the centuries and, even if not fully understood, at least assimilated.

Imagine that the manuscript of an unknown masterpiece by Dante, Shakespeare or Goethe is discovered in a monastery, an attic or a cellar. What would be the use, in this case, of beginning by setting up a critical apparatus? Surely the most important thing would be to make the work accessible to the general public. Once the specialists and the general public have grasped the essentials of the work, the task of the editor and of the annotator will be facilitated. Let us have no illusions, however. What is exceptional in literature – namely, the discovery of an important unknown text – is the rule in science, where very few of the outstanding works are known and appreciated by readers today. It is in their final version that scientific works are most easily understood and that the author's ideas emerge most clearly.

Lastly, an observation relating to the practical aspects of the edition is in order. Two aspects of the work involved in making texts accessible have been mentioned: their assembly, reproduction and arrangement, and their annotation. Whereas the former activity remains objective, the latter is clearly more subjective. It is therefore important to keep the two apart and not to introduce annotations into the original texts, which should be reproduced as historical documents with none but the unavoidable changes. Footnotes should contain only biographical or bibliographical information and explanations of symbols or vocabulary essential to the understanding of the text. The annotator's comments and observations may be grouped together in an appendix to the work.

The third stage consists of finding editors who are specialists in the subject dealt with in the group of texts they will have to annotate, which means that they must have worked actively on the texts. It is for this reason too that the texts have to be classified according to current criteria. Any other method of classification would increase the number of annotators working on the same volume. This is not practical and might lead to additional delays in publication. That is not the only factor which complicates this often very exciting stage. First of all the editor must be interested in the history of science and must know, or at least be able to read, the languages used by the author: Latin, in particular. It must be admitted that the number of scientists who know Latin is steadily decreasing.

Any one volume of an edition may of course contain several groups of texts on different subjects and therefore be annotated by different authors of various nationalities. As a result, it usually has to be printed in several languages. The present-day reader should not be put off by a certain heterogeneity. On the contrary, it should be welcomed as proof of true international collaboration.

It might be added that the role of the historian is very different from that of the scientist (as actor in the field of the natural sciences) and calls for certain practical skills which the scientist usually lacks and has to acquire the hard way.

The principles for publishing correspondence diverge here because the historical difficulties often take precedence over the scientific. The collaboration of professional historians or of scientists experienced in this work is often essential.

Lastly, the publication must be financed. Only a few aspects are mentioned here since this stage varies considerably from one case to another.

Human resources

In all cases qualified staff must be paid. This problem would be much easier to solve if more universities possessed institutes of science history. Collaboration in projects of this kind provides invaluable training, whether for students preparing theses or for assistant lecturers, and this type of institute can also provide the secretariat required. The work involved here is not that of the annotator, but the preparation of texts and manuscripts for the printer, the maintaining of liaison between the annotator, the publisher and the printer, the establishing of the principles of the edition, and the monitoring of their application. The tasks are numerous and require qualified researchers or small groups attached to a scientific institute, with which close contact is essential.

Material resources

Undertaking the publication of an author's complete works would be unthinkable today without the tools of modern technology such as photocopying machines, word processors, computers, and so on.

Financial resources

Apart from personnel and administrative costs, it is those of printing which make these editions expensive. The number of institutions or individuals able to purchase these volumes is not large enough to make the editions profitable. Fortunately, government bodies are showing more and more readiness to take their share of responsibility. However, without private sponsorship (through foundations, for example) such undertakings are not usually possible.

Means of distribution

The edition may be entrusted to a publisher with the necessary intellectual, artistic and financial resources. The publishing house must show great understanding and one cannot emphasize enough the merit of those publishers who are prepared to put energy into accomplishing this task. One might mention as an example the *Opera Omnia* of Euler, now in the hands of a third publisher and managed by a fourth generation of science editors.

Artistic resources are of the first importance, since not only do they do a great deal to make a book attractive, but they also can and indeed must increase the clarity of its presentation. Taking ancient texts written in unfamiliar language and using a formal style quite different from our own and making them accessible to a modern scientist is an arduous task which can only be achieved through long and close collaboration between graphic artists and scientists, and which goes far beyond the scope of the complete edition. Moreover, the possibilities opened by computers will certainly lead publishers to reorganize, though this is beyond the scope of the present article.

As for financial resources, there is little more to add. The reader will have understood that this type of undertaking is long and exacting, extending over decades and involving several generations of participants, and that continuity is a problem.

The number of difficulties to be overcome before the means of publishing these editions are found should not prevent us from taking on the task and making every effort to complete it, for the great scientific works are part of the world heritage and it is our duty to hand them down to the researchers of the future.

The History of Science at the Crossroads of
the Pathways towards Philosophy and
History[1]

To Jean Ladrière

Introduction

The history of science has one particular characteristic: science progresses. Whereas in the history of peoples and civilizations, high points are always followed by low points, and human progress winds its way through declines and disasters, as the past century shows only too well, for its part the development of science seems directed toward a goal. Of course, scientific progress can slow down and once a science is acquired, it can even fall into oblivion, but the important point is that the researcher is conscious of the fact that he is proceeding with the other scientists along the same pathway, though one whose direction is not yet fixed. Though it may be difficult, perhaps even impossible, for them to describe this direction clearly, they all have the certainty that it leads to a goal. Even the historian observing the results of this development over the course of a few centuries perceives not just an evolution, but an evolution towards a goal. Scientists have almost always been unanimous in this opinion, which is, I believe, justified, as long as one strictly abstains from adding the idea of moral progress into the mix.

But what does this progress consist in? On this matter opinions diverge. Personally I think that for the sciences called "exact," progress is manifest in the unification of theories, that is to say, in the fusions of branches that had been previously considered separate. Thus, hydrostatics and hydraulics became conjoined as hydrodynamics in the work of Daniel Bernoulli; hydrodynamics and aerodynamics found a common basis in the works of d'Alembert and Euler. At any rate, this process provides a criteria for defining progress and can even serve to measure it.

But these processes of unification or synthesis lead to a radically unhistoric dimension: the new laws discovered are valid everywhere and always. This process is not therefore merely a simple historic event; as I will try to show, it takes on, through the sciences, a philosophical significance.

For this reason, the historian of science, perhaps more than the historian of any other human activity, finds himself simultaneously confronted with two opposing obligations: he must study the progress of science in its historic development, and at the same time, through the sciences themselves, he must confront philosophy. On one hand, he has to follow and sometimes reconstruct the historic events in their richness

---

[1] Originally published as "L'histoire des sciences au carrefour du chemin vers la philosophie et des routes vers l'histoire," *Revue philosophique de Louvain* **94**, 3 (August 1996): 471–501. (Louvain-la-Neuve: Editions de l'Institut Supérieur de Philosophie).

as well as in their complexities. On the other, in order to show what progress owes to the discoveries of new laws of nature, the historian must have an idea of the particular science concerned, and this idea is located on quite another plane than the one of history: the laws of nature never submit to the game of history. But the first word in "law of nature" is "law," an idea that eludes science itself. Clarifying this idea, central to science as a whole, is a philosophical task. The result is that the historian is always confronted with philosophical questions. In what follows, I will briefly sketch the nature of these questions and compare them to problems that are properly historic.

This discussion comprises the following sections:

Science versus history
The scientific approach
The structure of science: the laws
Science and philosophy
The philosophy of science
Science and history
The pathways towards the history of science: historiography
Historiography and art.

Science versus history

"History, there's the enemy!" I would be tempted to put at the beginning of this discussion. That is to say, the enemy of a transparent and structured teaching of science, and thus of a teaching that doesn't present physics and chemistry as a heap of supposed laws and regularities, discoveries, it is said, with the help of observations or direct experiences, all disconnected from each other, without internal ties. A reasoned teaching of physics makes these internal ties evident; shows how some findings in different domains are based on the same theorems; shows how these theorems (for example, Green's theorem) play analogous roles in each of the domains and thus that some branches of physics have very similar structures; carefully distinguishes between fundamental laws and inferred *ad hoc* laws. Finally, only reasoned teaching is capable of giving sufficient importance to the introduction of fundamental notions and concepts.

As for this last point, alas, I too must confess: Father (or rather, considering my age, sons and daughters) *peccavi cogitatione et verbo!* I have sinned!

These deficiencies are largely due to the fact that often, without making a conscious effort, we, the teachers, merely follow tradition and recount what we want to teach in chronological order, that is, the order in which the first books addressed the matter at hand.

How many courses on quantum mechanics have been taught in this manner? One simply begins with what is the most difficult to comprehend, Planck's formula, because this is what provided the start for its development. One then continues with Einstein: the student is capable of understanding the photoelectric effect and the photon, but can he grasp its impact? Then comes Bohr's theory, which has been obsolete for decades. And when later the student learns the theory that is still valid today, some of the essential points will necessarily elude him. And I could go on.

It is evidently not necessary to begin with what was found at the beginning (and again, an agreement has to be reached regarding the question of what actually defines this beginning), but rather with what is *simplest.* Galileo, Newton (an author who was not embarrassed to be very difficult) and Euler knew this well. But it is unfortunately true that one doesn't recognize what is really simple without sufficient hindsight; in the higher-level courses this kind of process is therefore unavoidable. This is the reason why I believe that one of Dirac's claims to fame, as great as his discoveries, is that he, following Weyl, was the first to present quantum mechanics in an entirely systematic manner.[2] But this work of digestion, if I may use this expression, cost him almost thirty years of labor! I believe that every professor should constantly try to provide his students with the benefit of his own intellectual digestion, instead of, for example, following the *Berkeley Lectures on Physics*.[3] It is precisely this challenge that makes a teacher's task interesting. I shall return to this later.

In the texts destined for teaching, history must be relegated to the footnotes and to the bibliography, or, if necessary, to an appendix, in order not to interrupt the systematic presentation.

What I have just said concerns the relationship between science and history from the viewpoint of teaching and, more generally, communicating the sciences. But what is naturally of greater interest to the scientists is research, where this relationship becomes even more problematic. I doubt that there are many researchers who allow themselves to be inspired by the history of their particular branch of science. Of course, there are many mathematicians who have studied some of the works written long before their own, but then such works are considered as belonging to a present with which they are interacting. The past is quite another thing; our relationship with it is different. I will come back to this in a later part of this paper.

So, the reader will say to me, "Didn't we hear you say that you dedicated a large portion of your time to the history of science, yet gave no account of it to your students? How then, most esteemed Professor, do you have the nerve to call history the enemy?" My answer, and my praise of the history of science, could easily be the subject of several lectures, and further on will I take the liberty of saying some words about it. For the meantime, what is important to me is to underline the fact that the relationship between science and history is problematic.

The scientific approach

To begin I would like to give a simple and concrete example drawn from the activity of a physicist or a chemist, to give an idea of what the scientific approach effectively consists in.

Take a new material, a synthetic plastic for example, from which one would like to make the awnings or ropes. In order to do so it is necessary to know its elasticity, and thus it is necessary to measure it. One attaches the upper extremity of a plastic strip of length l to a frame and attaches a weight F to its lower extremity; one then measures the elongation of the strip caused by the weight. According to Hooke's law, Δl is proportional to the weight:

$$\Delta l = kF$$

where k is a constant. But the physicist taking this measurement knows, as Jacob Bernoulli showed, that k is not exactly a constant and thus for the large elongations he would do better to use:

$$\frac{\Delta l}{l} = f(P),$$

where P is Stevin's pressure: $P \equiv \dfrac{F}{A}$, where A is the section of the strip.

---

[2] See P.A.M. Dirac, *The Principles of Quantum Mechanics* (Oxford: Clarendon Press). The first edition was 1930, the second 1935, the third, 1947, and the fourth 1958.
[3] C. Kittel, W.D. Knight, M. A. Ruderman, *Berkeley Physics Course* (New York-San Francisco-Toronto-London: MacGrawHill, 1965).

However, one measurement alone is not sufficient to be able to apply this more general formula; in order to do that several measurements are necessary. Then all points measured must be interpolated with the help of a curve, the shape of which must be conjectured. This means that it is necessary to conjecture a constitutive law of the new plastic (since no one has yet calculated it beginning with its atomic structure) and to accept it as a temporary model. Instead of a synthetic product, we could also have taken a product called "natural," such as a precisely-defined alloy, so that the measurement is reproducible.

All of this appears simple, and indeed is certainly not very complicated, on the condition, however, that one can draw on the whole intellectual toolkit placed at our disposal by the development of geometry, physics and, in this case, chemistry. In fact, since we are dealing with elongation we need:

- the basic notions of Euclidean geometry;
- the basic notions of statics, in order to have the notion of force;
- a definition (physical and chemical) of what is meant by "homogeneous and simple material";
- a definition of "the elasticity of a body."

It is precisely for these reasons that for a long time there was such great confusion regarding the notions of elasticity and pressure.

It is important to realize that the intellectual approach sketched here is always two-fold. On the one hand, we construct a conceptual structure, we define the concepts that may be the subject of mathematical operations, in the framework of vectorial calculus, of infinitesimal calculus, etc.; on the other hand – and it is here that physics becomes a science of nature – we must provide, for each of these notions, a prescription for how to measure it. Only such a prescription provides the concept with its physical meaning, and only the logical rigor of the definition makes the reproducibility of the measurement possible. To measure, therefore, simply means to put one fact in relation to another fact, that is, to establish a connection.

It is only by means of such formal devices that modern science can and does speak of "law of nature." By "law of nature" I finally named the fundamental concept, or perhaps better, the "fundamental idea" that it is the goal of the scientific approach called fundamental, to know: *the discovery of the laws of nature.*

I will come back to these three notions, or these three ideas: law, nature, discovery. For the moment, I will only note that these three ideas are all very abstract and far from what is called "the experience of the senses." Each of us knows that defining one of these three notions, and *a fortiori* of all three, except by the enumeration of examples, is a very difficult task. As we know, examples are illuminating, but don't solve the problem.

The structure of science: the laws

"Laws of nature" therefore exist. However, the first word of "law of nature" is "law." There are many kinds of laws: legal, moral, grammatic, aesthetic, etc. The laws of so-called "inorganic" nature are formulated by mathematics and by logic. As far as mathematical laws are concerned, we all know more or less clearly what we are talking about, although the origin of these laws, the exact role that they play in the scientific approach, and the relationships between them and empirical observation are subject to discussion.

Before coming back to our main discussion, I must say a word about what I call "logic." Here I do not mean formal logic, nor logistics, which, for its part, is a part of mathematics. Rather, I have in mind an approach like the one previously considered by Plato, and which Fichte and Hegel tried to formulate systematically. In this approach one studies the genesis of the concepts and their significance; the role that they play, that is, how they are combined, how they are connected; how they enter into our judgments, our conclusions and, especially, how they apply to a science. Hegel added a new aspect to these investigations, where he speaks of "mechanism," of "chemism," and of "organism," which I believe corresponds to the study of what we call today a "model" and of which I have already given an example.

But let's first go back to mathematics, since modern physics, classic or quantum, as well as chemistry, is not conceivable without it. It is necessary to realize that the mathematician's approach to geometry and that of the physicist are not the same. For the mathematician there is a beautiful array of geometries: Euclidean, conformal, projective, Riemannian, with an antisymmetric metric, differential, topological, etc. The mathematician can formalize them and can make a science of them in which, to paraphrase Russell, "one never knows what one is talking about." In the words of a living mathematician, Enrico Bombieri:

> Mathematics is the study of relationships among objects. The universality of mathematics lies in the fact that the object itself is irrelevant—what really matters is the structure of the relationship. In this world of ideas, a mathematician is an explorer, an architect, an artist moving freely around in his never-ending quest for knowledge and vision with others, and seeing his discoveries put to good use.[4]

In other words, mathematics is an artistic construction that the physicist can apply according to his needs. However, I should add that, in contrast to Bombieri, I am less than sure than the object itself is never important for mathematics. An example that comes to my mind is the natural numbers, which constitute an unique object of investigation. But this is unimportant here, because what Bombieri said is certainly true for geometry.

On the other hand, the physicist, like the Greek geometer before him, is interested in an object to which he applies the beautiful things that he learned from the mathematician. While exploring this object by means of mathematics, he must justify this application. This is done, as we know, through the verification of one or several predictions.

This process can sometimes appear circular. The physicists usually react to these difficulties in two ways, one as aberrant as the other. The coarsest manner consists in inviting the student to observe as carefully and intensely as possible an experiment, while telling to him that the "proof" of the formula written to the blackboard will then appear obvious to him. Apart from the coarseness of the process, which, incidentally, works only for students whose brains are already sufficiently deformed by a large number of exams, one cannot see clearly by what process of cerebral distillation the formulae can be composed. The other method, slightly more refined, consists in bringing into play, against all logic, the supposed inductive proofs at the expense of the deductive proofs. In both cases one loses precisely one of the most precious qualities, and the one most fundamental for science: knowing the connection between different affirmations stated by a theory. Scientific theories are the networks by which the connections formulated by mathematics are confirmed by experiment.

To clarify this point I would like to choose as an example the application of geometry to the object in which we all live and move, which we commonly call

---

[4] Institute for Advanced Study, Princeton, Report for the Academic Year 1992–1993, p. 16.

"space." The history of the exploration of space teaches us that after numerous attempts, physics decided to describe space by applying the Euclidean geometry that had been developed by first studying the extension of bodies in space, and then by studying the space *itself.* Practical and aesthetic necessities were the bases of this research; later, the interest aroused by this research itself led to the study of the connections between these results. The Euclidean model was so successful that some philosophers, such as Kant, believed that its validity could be defended "a priori," that is, regardless of experience. Nothing of the sort: far from being obvious and trivial, this validity is quite limited. We know today that in the vicinity of the sun, and to an even greater extent, in the vicinity of a pulsar, Euclidean geometry is not applicable.

It is also clear that in this case there is no need to speak of a scientific revolution, but merely of a restriction on the applicability of the *ancien régime*. We can indicate clear and coherent numeric limits of the applicability of this régime, and thus, the application of Euclidean geometry need not be considered obsolete. The only thing that has happened is that its applicability has been restricted.

What I have just said about geometry is also valid for mechanics. In fact, I have already spoken of it, perhaps without anyone noticing it. In my mention of the different geometries I named one provided with an "antisymmetric metric." However, we have known for some decades (the central ideas actually go back up to Lagrange and Poisson), that the most powerful way to formulate classic mechanics consists in using simplex geometry, which is one characterized by an antisymmetric metric. Thus there is no fundamental difference between geometry and mechanics, nor even between geometry and physics: at every turn we encounter a dual approach. One has on the one hand a formalism, which, taken on its own, does not say "that of which one speaks," and on the other, an application that is specified by prescriptions that say how every quantity used in the equations must be measured. Because of the dual nature of the approach, I am happy to speak of a "Janus" in honor of Arthur Koestler, who noticed its importance.

The formal approach – the deduction and the proof starting from axioms – assures the consistency, or, if you will, the truth of the science. Further, applications that use univocal prescriptions for measuring a quantity provide us with assurance that one knows what one is talking about. That to which it is applied is sometimes called "nature."

However, the difference between formalism and application mustn't be confused with the difference between theory and experiment, because all scientific theory is conceived in view of an application, be it only as a model, as the scientist envisions his experiment against the backdrop of, and most often even bases it on, a rigorously formulated theory.

Science and philosophy

I know that many scientists today, as did many before them, say, "Philosophy, there is the enemy!" In saying this, the scientists' main criticism is that the philosophers don't pay enough attention to observation and empiricism, and that they overestimate the rational power of the human intellect. Ernst Mach was not alone in his verdict, which was inspired by the skepticism of Hume. Mach's criticism represents an extreme, but in some way reflects the attitude of the great majority of those who ask themselves how philosophy can be useful to science. This attitude was certainly not the one taken by Galileo, nor by Descartes or Leibniz. Nor was it the stance of Newton, Euler or Riemann, nor that of Einstein, Heisenberg or Pauli, and to that list I could add others. Personally, long before I was capable of reading these scientists, I had a strong distrust for Mach's ideas and those who thought like him, because they underestimated the role that mathematics play in the sciences. My initial idea became increasingly a certainty. Since the beginning of systematic research in geometry – I am speaking of the geometry that studies the space in which we live – one finds a number of philosophers involved: Thales (an empirical philosopher), Pythagoras, Plato and others. In short, I could never see the least progress in the transition from the metaphysical era to the positivist, as Mach said.

Let's hear what the scientist can expect from philosophy. I don't think that science must inevitably be based on philosophical foundations. It can always work independently. I recall, for example, the arrogance of a young so-called philosopher who reproached me, saying that there had not been any beautiful, major discoveries in physics in recent years and that it would have been a different story had the physicists studied philosophy seriously!

I am not thinking of a completion of science in the sense that, let's say, Sunday completes the workdays. But above all, I absolutely don't subscribe to the self-importance of some authors who pretend to be philosophers of the history of science, and who seem to want to say, "scientists find some results, but it is thanks to us that one learns to appreciate them." *Non ragionam di lor*,[5] as Dante said; it is labour lost. Rather, the usefulness of the philosophical approach for the sciences appears – note that I don't say "has been demonstrated" – through the scientific approach itself. Let us therefore examine the physicist's approach again.

The physicist often discusses the significance of his science and his activity with his colleagues only to fight for funding or to defend the thesis of a doctoral candidate that his colleagues judge to be, let's say, a little extravagant, not to say marginal, to present-day science. Such a defense is meant to be, and often is, a reasonable discourse, that is, consistent, logical and coherent. But its content certainly doesn't have anything to do with the science itself; it is clearly another kind of talk.

Let's remember that when we reason about science, we don't take a scientific approach. The approach taken can nevertheless be carefully considered and intellectually respectable. For example, Mach's anti-metaphysical position was not a scientific position; it was, let's be frank, the personal philosophy of Ernst Mach. And I underline "personal," because Mach didn't develop it in a systematic way, a thing that he would have declared to be impossible. Thus in my estimation it is not a philosophy about which one can actually enter into a dialogue.

Before going any further, I must open a parenthesis concerning mathematics, because here the mathematician is somewhat privileged. When he reasons about mathematics, using, among other things, abstraction, he often remains in the domain of mathematics, and never arrives to the science of nature. This aspect distinguishes mathematics from the other sciences and is in a sense a sign of superiority. But just the same, even the mathematician is not always able to remain inside mathematics, contrary to what I believed for a long time, and I will show why.

If we examine our scientific discourse more carefully, we can easily see that we constantly use notions that are not scientific at all. Examples of such notions are "true," "false," and thus "truth," "error," "reality," "clarity" (as in the statement "this explanation is clear"), "mechanism," "model," "reproducible" (as in "an experiment must be reproducible"), "cause," "causality," "matter," even "life," and many others.

None of these notions is represented in an equation, nor in a prescription making it possible to carry out an experiment, yet we constantly use them when we debate a scientific question. We do this precisely when we discuss our ideas about the

---

[5] Editor's note. Dante, *Inferno*, Canto III, verse 51.

relationships between the different branches of science and the question of their interdependence, as in the relationship of chemistry to physics. If we observe carefully, we can see that it would be extremely difficult to avoid these notions, and that if we were to do so, we would lose something of what is totally essential. These notions are philosophical notions, and we are all, like Molière's Jourdain,[6] doing philosophy without knowing it, except that, like Mach, we don't do it in a systematic manner.

Given mathematics' privileged position, mentioned above, one might hope to solve these difficulties by calling on mathematics. And indeed, this does make our life a lot simpler, which is precisely why mathematics was developed. Mathematics' capacity for abstraction with regard to nature is very powerful and certainly far from being exhausted. So one could hope that by examining the algebras of von Neumann and the non-commutative geometry that derived from it, it would be possible to arrive at a metatheory and thus avoid philosophy. But what we have seen shows that this is not possible. Physics progresses by leaning on theory and experiment, by tying mathematics to the empiric, that is, to what is accessible to us through the senses. The prescriptions for carrying out an experiment of course draw on mathematics, but the approach itself is not mathematical. A metatheory that only takes mathematical structures into account cannot therefore achieve what we are after.

I would like now to set forth an argument in favor of the necessity of philosophy which was proposed to me by a mathematician and philosopher friend. The merit of this argument is that it leads us to the heart of natural philosophy and of science. Here it is.

One concept that is as central for the mathematician as it is for the physicist, but which is certainly not a mathematical concept, is that of "law." But even the mathematician can only show us by an example what a law is. Using only the resources he has, he is unable to explain this concept, that is, to deduce its significance within the context of his own notions. For the physicist, the chemist and other scientists, as we saw, the task is even more complicated, since they work in addition with the prescriptions making it possible to carry out a reproducible experiment. One can also qualify them as laws, but evidently these are laws of another species: they regard not only our intellect but our actions as well.

Thus it is clear that, without a small amount of philosophy we are unable to even ask the question that concerns the foundation of all scientific activity – that is, "what is a law of nature?" or, perhaps more prudently, "what do we mean when we speak of law of nature?" – because, again, the ultimate goal of scientific research is the discovery of the laws of nature.

The philosophy of science

As I am not a professional philosopher, I must limit myself to a few non-systematic remarks. It goes without saying that here I won't speak of the philosophy of religion, nor that of art, state, politics, nor of history and especially not of morals, but solely of the philosophy of science. However, it has quite often been my experience that the comparisons between the philosophy of art and the philosophy of science is revealing, and is especially stimulating for science, in the same way that the history of science can benefit from the comparison of its goals and methods with those of the history of art.

---

[6] Editor's note. In Act Two, Scene Four of Molière's *Le Bourgeois Gentilhomme*, Mr. Jourdain learns from a philosophy master that he has been speaking prose all along without knowing it.

I will begin first by saying what philosophy is not:

Philosophy is not a science akin to the others: it does not explore a particular domain. Generally speaking, it does not study just a reality, not even a superior reality.

Philosophy is not a description or knowledge of some thing. It is not therefore, or at least not in the first place, an ontology or an understanding of a "being." On that head I am in disagreement with the line that goes from Aristotle to Saint Thomas, Descartes and Husserl.

However, philosophy must not be reduced to a demarcation of the terrain where science has to work and to an explanation of the manner in which we arrive to our knowledge. Here I am in disagreement with Kant and the neo-Kantians. Kant's work is a point of reference and very useful – even indispensable – for orientation (the works of Cassirer, among others, are proof of this). But the "table of the judgments"[7] is too narrow a starting point for a critique of science. It must be replaced, for this purpose at least, by mathematics in their infinite extent, that is, by those mathematics of which we know only a small fraction today. I said "at least," because we will see that something else must be added to it.

Be that as it may, Hegel was right to say that the philosopher who requires an epistemology *a priori* is "like the man who refuses to venture into water before he knows how to swim."[8] The epistemology must also be guided by experience. Therefore it can neither follow nor precede the work of the philosopher, but must perpetually go in parallel with it.

Philosophy, like science, operates on connections. I have the impression that this is almost the only thing on which all philosophers agree. It connects various notions with the help of judgments and conclusions, and sometimes with the help of "systems." These systems play something of the role in philosophy that models play in science, except that in general scientists are more prudent and more specific about the domain in which their models are applicable.

Philosophy does not concern itself with connections between phenomena and objects, but rather with objects and subjects that reflect on these objects. It also, on the one hand, concerns connections between different theories and branches of science, and those who attempt to understand them, and, on the other, evaluates their scope and their significance in life and in our existence. As an example, take the following question: "What is the significance of a particular piece of information about the oldest events of the universe, and therefore of what one commonly calls the 'origin of the world'?" This is first of all, of course, a scientific question, and must it be treated as such, but it is difficult to separate it from the rest of our knowledge and thoughts. It is therefore also necessarily philosophical.

How then must we proceed?

To arrive at an understanding of science, of the role that it plays in the whole *corpus* of our ideas and in our thoughts about the world, and of the appropriate position that it must occupy in life, we "must think about the scientist's thoughts." For example, we have to ask ourselves, "What makes it possible for science to organize phenomena (that is, relate them with the help of mathematical theories), and how is this done?" Or, "Why isn't it sufficient to merely write down the results of our measurements in a textbook, where they would peaceably coexist, without any internal connections, where to the contrary we can see that the results of our research

---

[7] Editor's note. See Kant's *Critique of Pure Reason.*

[8] Editor's note. See G. W. F. Hegel, *Encyclopedia,* William Wallace, trans. (2nd ed. Oxford: Clarendon Press), §10. See also *The Encyclopedia of Logic* (Hackett Pub Co., 1991), p. 34.

form increasingly larger entities, a little like the pieces of a puzzle?" And further, "How can it come about that these entities – and therefore the theories – are logical and mathematical constructions of an indisputable beauty?" Einstein and Dirac especially insisted on the importance of this point.

We have seen that the scientific approach requires the constant use of philosophical notions. Thus it is impossible to find the answer to these questions without the guidance of philosophy. But what philosophy, and what role can it play?

I would like to illustrate this last point by means of a discussion between a physicist and a philosopher, which, at least in part, actually took place. There was a philosopher who maintained that the notion of an elementary particle (an electron, for example) was subject to the same criticism as the notion of the ether. The physicist answered, "This comparison is nonsense: the electron exists, while the ether, as Michelson's experiment and Einstein's new mechanics showed, does not." But then, what does the "existence" of the electron consist in? The answer to this question is provided by the theory of the electron. Today the physicist who wants to speak of the electron has a choice between at least three theories that teach us the following things by turns.

According to Schrödinger's equation, the electron is a point having a mass and a charge, and that possesses all qualities that we generally associate with a "material object." At no time does it stop displaying all these qualities: its existence is permanent. A modification of this theory, due to Pauli, further attributes to the electron a spin and a magnetic moment. This theory allows us to calculate, or at least to explain qualitatively, the immense *corpus* of experimental data offered to us by the atomic and molecular spectra, as well as by chemistry. However, this theory is valid only if the speed of the electron is small in relation to that of light. Otherwise, we must resort to a second theory due to Dirac.

According to this second theory, the electron can fall in a "hole." Such a hole is, in fact, a positively-charged electron, called a positron. "Falling in the hole" means that the electron and the positron mutually cancel each other out: their charges disappear, and the energy contained in their masses is liberated in the form of two or three photons. Inversely, two photons can produce an electron-positron pair. In this theory one finds that the statement "the electron exists" means something other than it did at first: for Dirac, the existence of the electron is not permanent.

The third theory about the electron, the most satisfactory and exact, is given by "the theory of quantized fields," or quantum electrodynamics (QED). In this theory, electron and positron, proton and antiproton intervene in an entirely symmetrical manner, but here an electron is only a "mode of excitation" in the electron-positron field. Here "existence" means something else yet again: it is the field and the field laws that exist in the first place, and the electron is only a particular manifestation of these. Contrary to a material object considered by classic physics, such as a stone or a house, it doesn't possess any individuality.

On the other hand, the electromagnetic ether of Euler-Faraday-Maxwell-Lorentz effectively disappeared: Einstein and Dirac transformed it into a quantified field. But Einstein himself recognized that nothing prevented one from speaking of a gravitational ether, since the metric field, which no one has yet succeeded in quantifying, is a nonlinear field.

As a conseqence, the physicist, instead of attacking the philosopher, should have answered, "I am very happy to debate the question of the existence of the ether, the electron, or if you want, the quark, the electromagnetic field and the laws that govern them, but first of all, my dear friend, you must tell me what you mean by 'existing' and 'existence.' For the time being these are only words; you must develop the notions, and explain me what they mean and how I must use them. It is up to the philosopher to explain these notions, because they are philosophical notions."

Indeed, the scientist must define the scientific notions in an adequate manner, but for his part, the philosopher must define the philosophical notions.

How is this to be taken? Everyone who uses the notion of existence thinks unconsciously of a vague criterion taylored to suit their own domain, and each believes that his domain is the most profound, the one that defines "real existence." It could hardly be otherwise, since each person "explores fundamental reality," and so forth. Physicists are a good, or if you will, a bad example. In this case, mathematicians are best able to show us how the philosopher must proceed. Many of the mathematical notions used by scientists have been found and elaborated during the course of the study of a particular domain, but they have ultimately come to be defined without regard to this domain by an entirely different application. If it were otherwise, how could the rigour of the approach be guaranteed?

I think that in logic and in philosophy, that is, in their examination of notions, it is necessary to proceed in the same way. Thus one should explain what one means by "exist" without using examples, since if one proceeds by examples, one never arrives at the precise explanation of this notion, and the discussion will go around in circles. But in philosophy, it is more difficult. The notions can only define each other mutually, one in relation to another: we know quite well that the notions of "relative" and "absolute" make sense only in relation to each other, then all that is absolute is always relative, and in a sense the relative becomes an absolute...

Morever, everyone knows the sophisms that result from juggling with notions, and of which one is often the victim in certain discussions about science. The point that must be understood here is that the explanation and the definition of notions must not come from the objects that one wants to fix with the help of these, but that, as in mathematics, they must be used *in abstracto*.

Thus one should start with the simplest notions of all, and with the help of these, construct the more complicated ones. Plato tried to do this in *Parmenides* and *Sophist*. Much later Fichte also worked in this direction, and still later so did Hegel, especially in his *Logic*, where he took it upon himself to systematize notions. But although he took a step in the right direction, a good part of what Hegel stated arrives to us buried under an impenetrable, and sometimes erroneous, verbiage. Nevertheless, it seems to me that without such an approach, achieved *in abstracto*, philosophy loses its autonomy and only follows science without actually helping it.

Scientists fear the influence of philosophy, and object that philosophy disregards the essentially empiric nature of science. I think that the scientists who think this are mistaken. The goal of philosophy cannot be to erect a system that is valid once and for all; rather, anyone working *in abstracto* – that is, before any application – knows (or should know) that the application of his abstract formalisms to a certain domain must be always justified *a posteriori*, that is, empirically; a little like how Einstein showed that even the application of geometry must be justified empirically. The true goal of philosophy is merely to organize our minds and our ideas, to render our language consistent and our actions coherent.

In the same way that mathematics establishes relationships between magnitudes, numbers, figures, operators etc., philosophy establishes some connections between notions, thus its role in relation to science is a unifying one. In this way philosophy will help science arrive at the unity of our understanding and knowledge. We are now going to turn to history, and the role it plays in relation to science. There we will encounter the opposite attitude: the historian searches for the diversity and the infinite richness that we observe around us.

Science and history

I said earlier that the goal of science was "the discovery of the laws of nature." These "laws" are formulated with the help of mathematical language and logic; "nature" specifies the object to which the "laws" must apply, that is, the world in which we live. But what does "discovery" mean?

In a book given to me by my friend Jacques Weyers, Richard Feynman wrote that the discovery of a law of nature is "like discovering America – you only do it once!"[9] In other words, discovery is a historic act, with all that that implies, including, among other things, the impossibility of replicating it. I want to specify immediately that the notion of discovery must be taken here in its broadest sense. In particular, it is necessary to include what I referred to earlier as "digestion," the possession of a profound understanding of a result, its insertion in a general theory, the clarification of our understanding as well as the invention of a better experimental device for measuring an effect. All these are first steps "done only once." One can therefore see that history lies at the very heart of scientific life and its progress, and that it is essential. As I mentioned at the beginning of this discussion, the knowledge of history is not indispensable to understanding science, but if one wants to locate science in the context of other human activities that have a historic dimension, then one cannot do without it. And it is appropriate that the historian's approach should start with someone's discovery of a law.

Bearing this in mind, we can ask, "How is it possible that through actions – which although not random are at the very least often conditioned by luck – we managed to grasp truths of an almost universal validity?" I don't want to go into the philosophical aspect of the question, but I would like to say something about its historic aspect: what were the conditions in which a discovery was made?

When we say "history of science," we first of all say "history." History concerns everything that has taken place in our universe and has left a trace, beginning with the purely intellectual, like the writings of the Greeks and their rediscovery, and going to the purely circumstantial and material, like apples falling from a tree. It goes from theology, like the cosmic ideas of Petrus Peregrinus and Kepler, to economics, like the requirements of Roman administration, or the resources of medieval economy, all the way to the present requirements for managing our sources of energy, to mention a topic that is very fashionable today. I am forced to use the term "like" in each of these instances; indeed, history, and especially the history of science, is made up of actions performed by individuals.

History is the domain of empiricism; some distort it and want to develop some synthesis, and some distort it while approaching it, in some way or another, with a philosophy, as in the case of Hegel. History is infinitely complex; we can observe it "philosophically," but in its entirety it does not fit any theory.

The scheme of some modern authors and their pupils, who are looking to find "revolutions" in all corners of the history of science, and want thereby to grasp the essence of scientific progress, is a typical example, although it is something of a farce. At the moment the idea of revolution is currently most visible, and certainly the loudest war-horse of historiography. Allow me to say what I think about this scheme by entering into the concrete historic details of a discovery of prime importance that changed considerably the course of the development of science. Anybody who speaks of "revolution" will certainly agree that the discovery of

---

[9] Richard Feynmann, *The Character of Physical Law* (New York: Modern Library, 1994), p. 166.

Coulomb's law[10] was revolutionary. How was this discovery made? Here, following Whittaker,[11] is a brief summary of it.

After John Michell discovered, in 1750, an analogous law valid for magnetic forces, Franz Aepinus, one of the best researchers of the day in this field, attempted to find the law of attraction between two electric charges, supposing it to be a central force, but in vain. After Aepinus, we find a list of scientists who contributed to the formulation and confirmation of this law. (It is necessary to note that the list may not be complete, because in history one can be rarely sure.)

In 1760, Daniel Bernoulli conjectured this law – he is apparently the first to do so – but he didn't succeed in confirming it experimentally. His conjecture was only published in 1777 by his student Abel Socin. In 1766, Benjamin Franklin informed his friend Joseph Priestley that he observed a remarkable fact: a charge placed inside a charged metallic container shows no effect of force. On December 21 of the same year Priestley repeated this experiment and confirmed the result. The following year he wrote about it in his great history of electricity:

> May we not infer from this experiment that the attraction of electricity is subject with the same laws with that of gravitation, and is therefore according to the square of the distances, since it is easily demonstrated that were the Earth in the form of a shell, a body in the inside of it would not be attracted to one side more than the other?"[12]

But curiously, according to Whittaker,[13] Priestley hesitated to consider his experiment as a satisfactory proof. In 1769, John Robison of Edinburgh found experimentally that repulsion decreases by $1/r^{2.06}$ whereas attraction seemed to decrease less rapidly than $1/r^2$. He thus concluded a law of approximately $1/r^2$. From 1785 to 1789 Charles Augustin Coulomb published a set of seven mémoires dedicated to these questions.[14] I am not personally able to judge, but Chen Ning Yang, who studied them, told me one day that he found Coulomb's experimental confirmation rather weak. However, it is Coulomb's work that brought the law to triumph.

This discovery was the result of a long process that lasted twenty-nine years, and was accompanied by progress that was made simultaneously in several directions. Let's note in passing that the next twenty-nine years saw not only the taking of the Bastille and the French Revolution, but at least six other coups and restorations, and the battle of Waterloo! In comparison, the length of the revolution due to Coulomb's law seems a little long. We should also note at this point that it is awkward in the history of science to concentrate solely on the question of the priority of a discovery,

---

[10] Editor's note. In scalar form, Coulomb's law states that the magnitude of the electrostatic force between two point electric charges is directly proportional to the product of the magnitudes of each of the charges and inversely proportional to the square of the distance between the two charges.

[11] Edmund Whittaker, *A History of the Theories of Aether and Electricity*, vol. I: *The Classical Theories* (New York, Humanities Press, 1973).

[12] See Joseph Priestley, *The History and Present State of Electricity, with Original Experiments*, 3rd ed. (1775), vol. II, pp. 374. See also Whittaker, pp. 50–51.

[13] See Whittaker, *Op. cit.*, p. 53.

[14] Editor's note. Charles Augustin Coulomb (1736–1806) published his mémoires one through three (*Premier Mémoire sur l'Electricité et le Magnétisme, Deuxieme Mémoire sur l'Electricité et le Magnétisme, Troisième Mémoire sur l'Electricité et le Magnétisme*) in 1785, the fourth in 1786, the fifth in 1787, the sixth in 1788, and the seventh in 1989, all in the *Histoire de l'Académie Royale des Sciences*.

as scientific textbooks do. The historic reality is less simple, and fortunately more interesting. In the case we have just considered, I agree with Whittaker, who identifies in Priestley's observation and in the conclusion that he draws from it the most significant turning point of this evolution.

It is also necessary to recall that the passage of Priestley that I have just mentioned is in the very last chapter of a two-volume work, on the next to last page, whereas today this law is usually presented at the beginning of the textbook. This inversion of the order, the real revolution if you will, is owed precisely to what I referred to earlier as "digestion," which certainly required several decades. Like the actual discovery, this too is subject to a historic process. Even in our own century, where the pace of scientific life has accelerated along with everything else, the most remarkable process of digestion, to comprehend the coherent formulation of quantum mechanics, cost Dirac the work of almost thirty years, from 1930, the year of the first edition, until 1958, the year of the fourth, and only the last one contains the quantization of fields viewed from within the same unified formalism. Dirac started over three times!

But the history of the discovery of Coulomb's law that I have presented here is not history in its entirety. It must also be observed that these same years also witnessed the beginning of another development.

In one of his *Letters to a German Princess*,[15] the letter CXXXVIII written on 20 June 1761, Euler writes:

> The subject which I am now going to recommend to your attention almost terrifies me. The variety it presents is immense, and the enumeration of facts serves rather to confound than to inform. The subject I mean is electricity, which for some time past has become an object of such importance in physics that everyone is supposed to be acquainted with its effects.[16]

He goes on to say,

> Natural philosophers ... are almost every day discovring new phenomena, the description of which would employ many hundreds of letters; nay, perhaps I should have never done. And here it is I am embarrassed."[17]

---

[15] Editor's note. Euler's *Lettres à une princesse d'Allemagne sur divers sujets de physique & de philosophie* were written approximately two a week from April 1760 to May 1763 as lessons in science for the Princess of Anhalt-Dessau. They were published for the first time in three volumes beginning in 1768. The translations quoted here are from the 1833 English edition by David Brewster.

[16] English translation by David Brewster, from *Letters of Euler on Different Subjects in Natural Philosophy Addressed to a German Princess*, 1833, Letter XXIII, vol. 2, pp. 79–80. The original text is: *La matiere sur laquelle je voudrois à présent entretenir V. A., me fait presque peur. La variété en est si surprenante et le dénombrement des faits sert, plutôt à nous éblouir qu'à nous éclairer. C'est de l'Electricité dont je parle, et qui depuis quelque tems est devenue un article si important dans la Physique, qu'il n'est presque plus permis à personne d'en ignorer les effets.*

[17] English translation by David Brewster, from *Letters of Euler on Different Subjects in Natural Philosophy Addressed to a German Princess*, 1833, Letter XXIII, vol. 2, p. 80. The original text is: *Tous les Physiciens en parlent aujourd'hui avec le plus grand empressement, et on y découvre presque tous les jours de nouveaux phénomenes, dont la seule description rempliroit plusieurs centaines de lettres; et peut-être ne finirois-je jamais. Voilà l'embarras où je me trouve.*

One needs only to open Priestley's book to verify the exactness of Euler's remark; how then does he proceed to pull it off?

After having described the most important phenomena that have been observed, in the following letter, which is entitled "The true Principle of Nature on which are founded all the Phenomena of Electricity," he writes:

> There is no room to doubt that we must look for the source of all the phenomena of electricity only in a certain fluid and subtile matter; but we have no need to go to the regions of imagination in quest of it. That subtile matter denominated *ether*, whose reality I have already endeavoured to demonstrate, is sufficient very naturally to explain all the surprising effects which electricity presents.

And, ever the optimist, Euler adds:

> I hope I shall be able to set this in so clear a light, that you shall be able to account for every electrical phenomenon, however strange an appearance it may assume. The great requisite is to have a thorough knowledge of the nature of ether.[18]

In other words, Euler proposes a presentation that is qualitative of course, but also systematic, that is, deduced from certain mechanical principles, and from all observed phenomena. What are these principles? These are simply the "Euler equations of hydrodynamics," found some years before, which he applies to a model viz. the hydrodynamic ether, or, as we would say today, scalar. Just one year before the writing of this letter to the princess, Euler had found the wave equation in three dimensions (today called the d'Alembert equation) which he proposed for describing optical phenomena. This discovery sustained his thesis that phenomena of electricity and optics were intimately linked by a common dynamics. Whittaker assures us that Euler was the first to propose this hypothesis.

Thanks to Euler's hypothesis, precisely as was the case a little later thanks to the law of Bernoulli-Priestley-Robison-Coulomb, the phenomena of electricity, which until that time had constituted an independent field, were linked to mechanics. So the new science came out of its isolation, which opened new pathways for later research. As a consequence, instead of one revolution, and this is what interests us here, we find two of them. The $1/r^2$ law marks the beginning of the development that would lead to the laws of Ampère, Wilhelm Weber and others, while at the same time Euler's idea provides the starting point for the physics of fields, first of all scalar, and then, long after Euler, tensorial. The most important reader of the *Letters* was Faraday, the greatest experimenter of his century, thanks to whom these two currents came together in the theory of Maxwell.

The reason I am saying all this is to show that these two scientific currents coexisted peacefully, even though any given physicist might have been a staunch

---

[18] English translation by David Brewster, from *Letters of Euler on Different Subjects in Natural Philosophy Addressed to a German Princess*, 1833, Letter XXIV, vol. 2, p. 82. The original text is: *Il n'y a aucun doute qu'il ne faille chercher la source de tous les phénomenes de l'électricité dans une certaine matiere fluide et subtile, mais nous n'avons pas besoin d'en feindre une dans notre imagination. Cette même matiere subtile qu'on nomme l'Ether, et dont j'ai déjà eu l'honneur de prouver la réalité à V.A., est suffisante pour expliquer très naturellement tous les effets étranges que nous observons dans l'électricité. J'espère mettre V.A. si bien au fait de cette matiere, qu'il ne restera plus aucun phénomène électrique, quelque bizarre qu'il puisse paroître, sur l'explication duquel Elle puisse être embarrassée. Il ne s'agit que de bien connaître la nature de l'éther.*

supporter of one and just as staunch an opponent of the other. These *two* currents both advanced science, and are therefore equally important for the historian. When one starts to speak of revolution, one refers to one of the two currents while necessarily disregarding the other, the supposedly non-revolutionary one, and depicts the sort of picture that is only a mere caricature of the past, which was in fact much richer. In short, the history of science operates in a completely different way from the history of nations. With hindsight, today we see clearly that the simple term "turning point" characterizes what happened better than the pompous term "revolution." This dual development also illustrates the point that the famous notion of "paradigm" is harmful for the historian: it only leads to unfortunate simplifications.

History – and the same can be said of history of science – doesn't appear to be governed by laws; no law of history has been discovered, at least not yet. As everybody can see, the progress of the science goes along first in one way, then in another. Think of Heisenberg and Schrödinger. The essence seems to rest always on an act that is a matter of the individual's productive imagination, or *productive Einbildungskraft* as Fichte called it.[19] In the case of Coulomb's law as in that of the ether, the role played by each of the actors shows this to us. But the productive act, taken in itself, eludes our analysis. We can note it, depending on the circumstance, we can approach it somewhat, but, alas, acts of imagination don't reproduce themselves on command. And the further away in time this act is from us, the less it is possible.

Alexandre Koyré, the author of the *Galileo Studies*,[20] elegantly formulated the consequences of the unique character of the creative act at the international symposium on the history of science held in Oxford in 1961. In response to a serious and respected American scholar who proposed a vast program intended to make it possible to explain all scientific discoveries beginning with their cultural environment, Koyré retorted, "we can explain adequately why science was not born and did not develop in Persia or China ... and although we may be able to explain why it was possible for it to be born and to develop in Greece, we still cannot explain why it did so in fact."[21] Faced with the facts, the historian – and in particular the historian of science – must define a point of pause, and at that point think twice before imagining easy and necessarily imperfect theories to explain them.

When he examines history, the philosopher must limit himself to asking questions. One might say that history is for philosophy what nature is for science. Constructions and scientific theories must find their confirmation in experiments; in the same way, the constructions of philosophers must answer to history.

But then, you will say, philosophy doesn't have anything to teach the historian. The question is beyond the scope of this article, but I would like all the same to make a remark. I think that the constructions of philosophers, "philosophies" if you will, have very often constituted an enrichment of our concepts of history. They developed, refined and perfected our concepts and notions about history, and above

---

[19] See Fichte, "Concerning the Difference Between the Spirit and the Letter in Philosophy, First Lecture: Concerning the Spirit and the Body as Such," pp. 192–199 in *Fichte: Early Philosophical Writings*, Daniel Breazeale, trans. and ed., (Ithaca, NY: Cornell University Press, 1988), especially p. 193.

[20] Alexandre Koyré, *Les études galiléennes* (Paris, Hermann, 1939); Eng. trans. *Galileo Studies*, John Mepham, trans. (Brighton: Harvester Press, 1978).

[21] See p. 855 in "Commentary" by Alexandre Koyré, in A. C. Crombie (ed.), *Scientific change. Historical studies in the intellectual, social and technical conditions for scientific discovery and technical invention from antiquity to the present. Symposium on the history of science, University of Oxford, 9–15 July 1961* (Heinemann, London, 1963), pp. 847–857.

all they added new dimensions to the space in which we imagine it and in which we reflect. The increased popularity of historiography in Germany after Herder, Goethe, Schiller and Hegel testifies to this.

The pathways towards the history of science: historiography

In the title of this article I wrote "pathways," in the plural, first of all because there are various sciences, each with its own history. Further, the history of each must be written by a scientist in the field, and here too there will also be more than one history, especially as these different histories often developed without a lot of contact between each other.

Whereas philosophy is, of its very nature, a unifier, history is diversity and infinite richness. The question therefore is not what needs to be done, but rather what one wants to do. In the wealth of documents, what gives the most satisfaction, and in what form can I communicate it so that it is equally satisfactory? Science is not therefore situated "between" philosophy and history, because these two approaches achieve their ends in radically different programs, and we arrive there by following orthogonal directions. The more one mixes philosophy and history, the more one misses the interesting things that history can show us. Here I want to list only some of the ways to approach history. The few modest remarks that follow are far from being exhaustive; one never exhausts history, not even the smallest part of it. But it might happen, and in fact I hope this is the case, that one or another of my readers will be attracted to the history of a science, and thus that one or another of these remarks will be useful.

1. One can study the history of a particular field or, more precisely, of a key problem, such as the history of the concept of the magnetic lines of force, or the history of elasticity in a given period. This constitutes one of the most enriching approaches for the author as well as for the reader, since it makes it possible to show how the discipline developed from its very beginning, or perhaps even before. In this kind of limited setting, the reader can easily follow almost all the factors that contributed to the progress of the discipline, and thus acquire an overall picture of it. The first of these two topics has been studied by Patricia Radelet-de Grave,[22] who introduces the reader to the sources and documents from several centuries that she was able to find. For the second, the reader who wants to spare no effort will find in Clifford A. Truesdell's voluminous history of elasticity[23] probably the most discerning and most complete story that we possess today of an entire field of physics.

In this type of approach, one runs the risk of disregarding the influence of neighboring domains: physicists, for example, risk disregarding the influence exerted by chemistry or astronomy. Patricia Radelet-de Grave did well to concentrate her attention first on a presentation of the original texts, and commenting on them later, so that each is directly connected to the past just as it was.

2. Another possibility, currently much in vogue, is provided by the biographies, on the condition that they follow Einstein's advice (which was the advice followed by Einstein's biographer, Abraham Pais), which was to the effect that the only really interesting things in the life of scientist are his works. There are some exceptions of course, like Copernicus and the years that he spent in Italy, and Kepler, not so much

---

[22] Patricia Radelet-de Grave, *Les lignes magnétiques du XIIIe siècle au milieu du XVIIIe siècle, Cahiers d'Histoire et de Philosophie des Sciences,* Nouvelle série, vol. l, Paris, 1982.
[23] Clifford A. Truesdell, *The Rational Mechanics of Flexible or Elastic Bodies, 1638–1788, Leonhardi Euleri Opera Omnia,* Series II, vol. 11b (Zurich: Orell Füssli, 1960).

because his life was adventurous, but rather because it was so heroic and exemplary. The portraits that Koestler has sketched for us in *The Sleepwalkers* are magnificent.[24] As for Einstein himself, Pais was capable of doing justice to the adversities that life held in store for the hero of his story as well as to his scientific triumphs.

On the other hand, a profound knowledge of Newton's *Principia* teaches us far more about its withdrawn and introverted author than many biographic details can, and the same thing is probably true for Jacob Bernoulli, who in many ways resembles Newton.

Reading a work allows us to see how a discovery emerges from an intellect, that is, how intelligence is formed. One can sometimes even see when, and how, the new idea was formed, how it was received, how it spread, and how it was finally absorbed by the scientific world.

It must, however, be underlined that numerous publications testify to the fact that their authors don't understand the difference between the history of scientists and the history of science, and that they are not very familiar with the latter.

3. One can also describe an epoch, either by showing a segment of time through a single branch, or by painting a picture of the whole scientific age. There is much to be learned from books such as those by Wolf[25] or Crombie[26], especially regarding long-ago eras, when the river of science was yet but a stream. This is true even when the author is unable to grasp everything he writes about, which is rarely possible, since each field should be presented by someone who has worked in it. Still, such books are often very useful as an introduction to a particular field.

4. There are above all the works dedicated to a particular question, most often published in a journal. Fortunately, it is easier to isolate a question in the field of history than in that of science, and to present it without too many ties to other problems. But, alas, this kind of work most often concerns the specialized historian and not the scientist who wishes to learn how the domain in which he works was formed. This aspect of the question is often neglected by the historian. In my opinion, the historian should write for scientists and engineers more often than he does.

We can see, therefore, that the scholar who occupies himself with history always makes a choice, exactly like the scientist who chooses his problem. I said earlier that to speak of history implies the citation of concrete examples, and this is also valid for historiography. I therefore choose the example that I know best, and take the liberty of indicating the reasons that led me to make this choice.

My choice, like all choices that we make, was not made in an absolute vacuum; it was predestined, if I may be permitted to use this term, by where I was born and where I went to school, as well as my family relationships. Like many others, I occupy myself with what has been done in my country, a quite frequent motive among the historians, and generally a happy one. To be precise, I have been called on three times to edit scientific works.

But, independent of these requests, I have always believed that to edit scientific works and, as I like to emphasize, to make them "accessible" are the central tasks of the historian of science. This idea of mine has been strengthened, over a long period

---

[24] Arthur Koestler, *The Sleepwalkers: A History of Man's Changing Vision of the Universe* (1st ed., London: Hutchinson, 1959).
[25] A. Wolf, *A History of Science, Technology and Philosophy in the Eighteenth Century* (1st ed., London: G. Allen & Unwin Ltd., 1935; Macmillan, New York 1939).
[26] A. C. Crombie, *Augustine to Galileo: The history of science AD 400–1650* (1st ed., London: Falcon Press, 1952). Reprinted as *The History of Science from Augustine to Galileo* (New York: Dover, 1996).

of time, by, among others, J. R. Oppenheimer, who put it succinctly: history is entirely in the documents.

I said earlier that the discovery of the laws of nature was the scientist's ultimate goal. However, the work in which such a discovery is published for the first time generally marks the closest point to this that it is possible for us to reach. Sometimes we might still possess the scientist's notes or a first draft. These elements can bring some indications, but they most often mark a stage at which the author didn't yet know clearly how to formulate his discovery, or perhaps had not yet even arrived at it. Thus it is much wiser to take the publication as a starting point for historic research, a little like the way the scientist himself does.

But what do we mean by "making an old work accessible to today's reader, and in particular, to today's scientist"?

We must first concentrate on the question of the significance of the particular discovery for the scientist today. Only after we answer this can we answer the second question, which regards the significance of this discovery for the scientist of the period. The answers to these two questions are generally very different. I expressly put them in this order, because the answer to the first, which is a purely scientific question, is situated outside of the historic discussions, and is therefore often easier to provide. Further, a reflection on the answer to the first question helps the historian to better comprehend the earlier scientist's situation. Thus, only a good knowledge of the hydrodynamics of today makes it possible for us to understand the works of Newton, Jacob Hermann and the Bernoullis in this field, to evaluate their achievements, and to grasp their attempts.

I would like to illustrate all this in the case of my own choice. Having found my place in the eighteenth century, I believe that it has turned out to be a good thing, since this period is always – and, I am tempted to say, fortunately – disregarded by the historians of physics. I like to say that it is entertaining to explore another country, but that it is even more entertaining to explore another continent. The science of the eighteenth century is in many ways very distant from our own. We find there, for example, with the exception of astronomy, that there are very few precise measurements.[27] Theory – and I myself am a theoretician – played the leading role in mechanics, and mechanics constituted at least three-quarters of physics. Mathematics were important for physics at a level that would only be equalled in the last quarter of the twentieth century. Therefore, although the physics of this period reveals another continent to us, it is in some ways closer to us than the one of the past century, which was often characterized by an unbridled empiricism. Thus the physics of the eighteenth century has a certain relevance for us. There, among other things, we find the first tentative formulation of a unified field theory. There is also a supplementary advantage: the documents that arrive to us from another continent provide a window on the world, and the university and its departments can never have enough of such beautiful windows. How then to publish these documents?

This is a big question, one that has caused and will continue to cause many a debate. Patricia Radelet-de Grave and I wrote a paper on these problems,[28] but here I

---

[27] Editor's note. David Speiser raised this issue at the 1961 Oxford symposium on the history of science. See A. C. Crombie (ed.), *Scientific change. Historical studies in the intellectual, social and technical conditions for scientific discovery and technical invention from antiquity to the present. Symposium on the history of science, University of Oxford, 9–15 July 1961* (Heinemann, London, 1963), pp. 489–490.

[28] See David Speiser and Patricia Radelet-de Grave, "Publishing Complete Works of the Great Scientists: An International Undertaking," pp. 117–125 in this present volume.

will limit myself to mentioning only a few points. The first concerns the question of what "make accessible" means. The answer is that it is necessary to publish works of the past in such way that today's scientist, as well as all those who are interested in their content, can read and understand these works without too many difficulties. In order to achieve this, it is necessary to present the formulas as today's reader is accustomed to seeing them, that is, to extract them from the current text, without, of course, modifying their substance. For the Bernoulli edition, Patricia Radelet-de Grave, and then assistant editor Martin Mattmüller, dedicated quite a bit of time to this problem. It is especially necessary that the introduction and the commentaries to these works be written by a scientist working today, because the historic authors are nearly all difficult to read (Euler is an exception to this rule).

The second point, tied to the first, is of a philosophical nature. It is necessary to radically reject the idea that historiography works in an absolute space, or, if you will, that an objective, unique and complete reconstruction of the past exists, one that can be written once and for all. On the contrary, a work on the science of the past is nothing but a relation between the past and the period in which the editor writes. To see that this is true, we need only reflect that a history of Newton's discoveries in mechanics written after 1905, the year of Einstein's discovery of the new mechanics, is necessarily very different from a history written some years earlier. It is therefore obvious that in just fifty years from now, and perhaps even earlier, the historian will view the past very differently; he will first of all be interested in other questions, and he will place an emphasis on other things than we do today. What I have said regarding Einstein's mechanics is also valid for the Bernoullis, Euler and the others. All that we can hope for is to edit the works of the ancients from the most advanced scientific viewpoint possible. The original texts maintain all their value, while on the other hand the introductions and commentaries will certainly be useful to later successors, just as we used the works of Nicolaus I Bernoulli, who published the works of Jacob Bernoulli, and Gabriel Cramer, who published the works of Johann I Bernoulli, even though their opinions about their achievements in mathematics, and especially in physics, are necessarily very different from our own.

Historiography and art

I would like to end by mentioning a point which is somewhat marginal to the chosen topic, but which is nevertheless of great importance for it; at the same time, what I have to say is a confession of a personal regret.

The presentation of history is also, and especially, a question of an artistic achievement, that is, of the historian's creative power. It is well known that the historians that we admire the most – Gibbon, Burckhardt, Fustel de Coulanges, Mommsen, Churchill – were also artists; it is not not for nothing that the last two received the Nobel prize for literature. It is the same for the history of science, and the regret that I must confess is that I don't have a gift for writing, which is indispensable in the history of science for depicting a great figure of the past or an entire period.

I would like to conclude by naming some historians of science who have distinguished themselves through their artistic gifts.

Among the ancients I would like to mention here the discussion of the two rival theories of gravitation, that of Descartes and that of Newton, by Johann I Bernoulli.

This discussion confirms what Mach said when he called Bernoulli "an aesthetic genius."[29]

Among the historians of the twentieth century, my first choice is the book by Arnold Sommerfeld, *Atombau und Spektrallinien*,[30] which is of course a scientific book, but Sommerfeld's artistic gifts are apparent when he presents the prehistory of his topic. Sommerfeld was inspired by the *Vorlesungen über die Entwicklung der Mathematik im 19. Jahrhundert*[31] by his teacher Felix Klein, who was in his turn, I suspect, inspired by, or at least wrote in the tradition of, Goethe's *Materialien zur Geschichte der Farbenlehre* (*Materials for a history of the theory of colors*).[32] I also take the liberty of adding the name of Andreas Speiser, who in his day knew how to open up history of sciences to new horizons. He was the first to draw the attention of scientists and historians to the mathematical knowledge of the ancients contained in their artistic creations, especially in the jewelry of the Egyptians and Arabs.[33] That which at the time might have appeared as a hobby, today constitutes a respected branch of the history of science.

I think that the great influence that Alexander Koyré exerted is due in large measure to the aesthetic fascination exerted by his *Galileo Studies*. The same holds for *The Sleepwalkers*, where Arthur Koestler succeeded in painting majestic portraits. That of Kepler is particularly praiseworthy.

I would also like to name Clifford Ambrose Truesdell, for the power of some of his writings, and André Weil for the intellectual light that shines in his history of number theory.[34] Abraham Pais combines an extraordinary knowledge of the physics of our century with an enviable gift for writing, a gift that we can also admire in the writings of I. Bernard Cohen, co-author with Alexandre Koyré of the critical edition of Newton's *Principia*.[35]

Why do I mention art, these authors and their books? Simply because these provide the best and most convincing answer to the question, "Why do we interest ourselves in the history of science?"

Translated from the French by Kim Williams

---

[29] Ernst Mach, *The Science of Mechanics*, Thomas J. McCormack, trans., 6th ed. (Chicago: Open Court Publishing, 2010), p. 522.

[30] First published in German (Braunschweig: Friedrich Vieweg und Sohn, 1919), published in English as *Atomic Structure and Spectral Lines*, from the 3rd German ed., Henry L. Brose, trans. (London: Methuen, 1923).

[31] First published in German (Berlin: Springer, 1926–1927), published in English as *Development of Mathematics in the 19th Century*, M. Ackerman, trans. (Brookline, MA: Math Sci Press, 1979).

[32] First published in German (Tübingen: G. C. Cotta, 1810). Republished in *Goethes Werke* (Munich: C.H. Beck, 1981).

[33] See Patricia Radelet-de Grave, "Andreas Speiser (1885–1978) et Hermann Weyl (1885–1955), scientifiques, historiens et philosophes des sciences," *Revue Philosophique de Louvain*, vol. 94 (August 1996), pp. 502–535.

[34] A. Weil, *Number Theory: An approach through history from Hammurapi to Legendre* (Basel: Birkhäuser, 1984).

[35] I. B. Cohen and A. Koyré, eds., *Isaac Newton's* Philosophiae naturalis principia mathematica. The third edition (1726) with variant readings, 2 vols. (Cambridge, MA: Harvard University Press, 1972).